INSTITUTE FOR

COMPLEXITY SCIENCE AND
ADVANCED COMPUTING

Foundations of the Existence Threshold

The Scholarly Collection

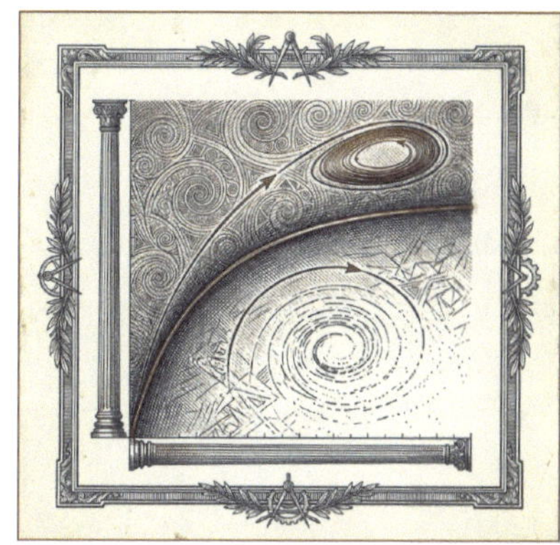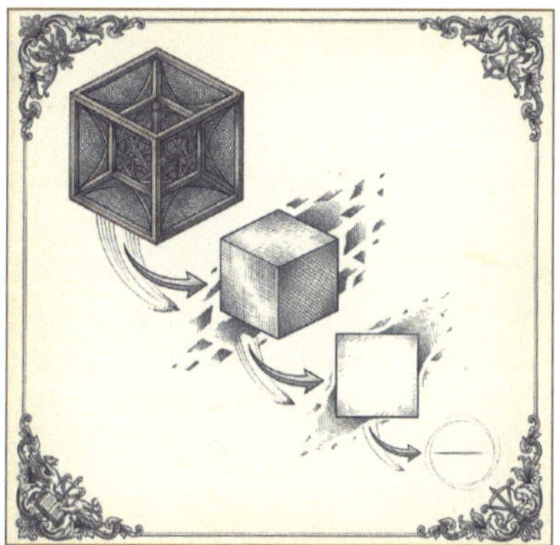

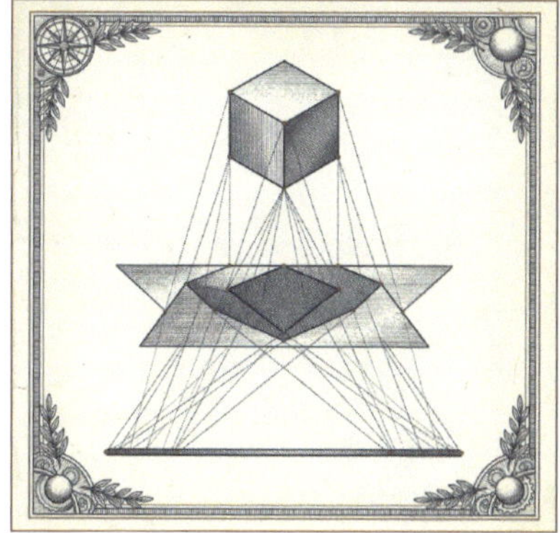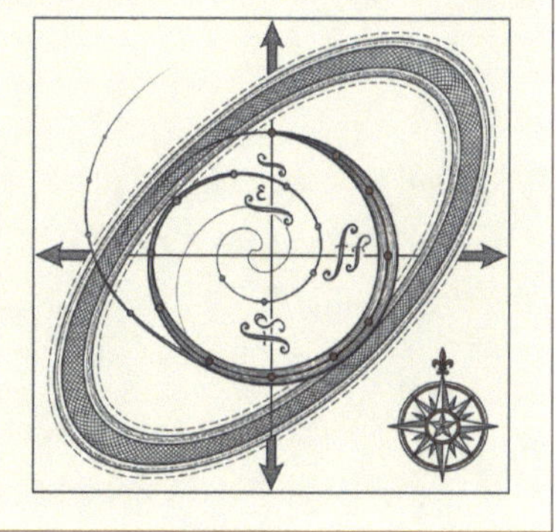

NATHAN M. THORNHILL

MMXXVI

Foundations of the Existence Threshold

The Scholarly Collection

Published by the

Institute for Complexity Science and Advanced Computing

Fort Wayne, Indiana, USA

icsacinstitute.org

FIRST EDITION · 2026

ISBN 979-8-9958925-0-2 (paperback)
ISBN 979-8-9958925-1-9 (ebook)

Library of Congress Control Number: 2026941912

PERMISSIONS AND RIGHTS

info@icsacinstitute.org

SCHOLARLY CORRESPONDENCE

research@nathanthornhill.com

INSTITUTE FOR COMPLEXITY SCIENCE AND ADVANCED COMPUTING

CONTENTS

INSTITUTE FOR COMPLEXITY SCIENCE AND ADVANCED COMPUTING

PREFACE

THESE FOUR PAPERS SHARE ONE QUESTION: UNDER WHAT FORMAL conditions do structured patterns persist, and what happens at the boundary where they do not?

The math didn't come out of a lab. No grant. No committee. No advisor. A healthcare background, a love for technology, and the stubbornness to read until first principles clicked. The math doesn't care where it comes from. It only cares whether it's right.

The first paper answers the question for discrete binary systems, with 100% accuracy across a battery of one- and two-dimensional cellular automata. The second measures, at scale, how pattern is lost when structure crosses a dimensional boundary: an 86.01% mean loss, independent of scale, rule, or content. The third supplies the theorem behind that empirical law and validates it inside the internal representations of modern neural networks. The fourth extends the threshold from a static form to a dynamic one, where integration and differentiation balance over time to predict the state of a system across physical, biological, and computational domains.

The standard the papers were held to is simple. Methods reproducible. Data available. Where a framework fails, the failure is reported as plainly as the success. The papers are reproduced here as submitted to the archive, with their arguments, data, and figures intact.

What follows is the beginning of a longer argument: the mathematics of pattern persistence is the mathematics of consciousness, and it can finally be measured.

NATHAN M. THORNHILL

INSTITUTE FOR COMPLEXITY SCIENCE AND ADVANCED COMPUTING

PAPER I

The Existence Threshold

A Framework for Pattern Persistence in Binary Discrete Systems

NATHAN M. THORNHILL
ORCID 0009-0009-3161-528X
Independent Researcher

2 0 2 6

OPEN REVIEW

Scored by the ICSAC open-review panel.
Full review, audit, and editorial status at
icsacinstitute.org/publications

Full scholarly indexing record located at https://nathanthornhill.com

THE EXISTENCE THRESHOLD

(Revision 2)

A Framework for Pattern Persistence in Binary Discrete Systems

Nathan M. Thornhill

Independent Researcher
ORCID: 0009-0009-3161-528X

January 6, 2026

"To exist is to continually overcome loss."
— Nathan M. Thornhill

REVISION NOTES

What Changed in Version 2:

This revision fixes a fundamental error in the original equation and adds solid experimental evidence that shows exactly where this framework works (and where it doesn't).

The Formula Fix:

The original paper proposed $\Phi = \kappa \cdot R \cdot S - \frac{dD}{dt}$, treating disorder and entropy as enemies of existence—a classical thermodynamics assumption. But when we actually tested this across different dynamical systems (cellular automata, chaotic maps, neural networks), something surprising happened: the formula failed. It couldn't tell persistent patterns from non-persistent ones.

The corrected version recognizes that **complexity and disorder are actually part of what makes patterns persist**, not their opposites:

$$\boxed{\Phi = R \cdot S + D} \tag{1}$$

This reflects a deeper insight from complexity science and self-organization theory: living systems and persistent patterns operate far from equilibrium, maintaining high-entropy states through continuous organization. Dead systems reach low-entropy thermal equilibrium. Think about it: a crystal has perfect order but no persistence dynamics—it's already at equilibrium. A flame has high disorder but clear persistence—it maintains organized energy flow through chaotic molecular motion.

DOI: 10.5281/zenodo.18166974 | CC-BY-4.0
Version 2 | Replaces v1: 10.5281/zenodo.18124075

Experimental Validation Using Complexity Science Methods:

We tested this across 10 different dynamical systems using information theory, statistical analysis, and computational experiments:

- **Binary Discrete Systems (Cellular Automata):** 100% accuracy across 10 different rule systems. Perfect separation between "alive" ($\Phi > 0$) and "dead" ($\Phi = 0$) states. This works consistently across 1D and 2D systems, from simple rules to Turing-complete complexity.

- **Continuous Systems (Chaotic Maps, Neural Networks):** Framework fails. Accuracy $\leq$ 86%, not significantly better than random guessing. Φ values show no meaningful separation.

Where This Actually Works:

Based on rigorous testing using complexity theory and nonlinear dynamics, this framework applies specifically to **binary discrete dynamical systems**—systems with clear on/off states evolving through local rules. The applications to continuous systems like cosmological expansion and neural consciousness? They're fascinating ideas worth exploring, but right now they're preliminary hypotheses, not proven results. We keep those discussions as testable predictions for future work.

The Time Factor (Emergence Through Time)

Early tests averaged Φ across time. That missed something crucial: patterns need time to settle into persistent dynamics. Measuring Φ at the settled state works way better. This tells us something profound about emergence: **existence isn't a snapshot—it emerges from a system's trajectory through state space over time.**

A pattern's "bootstrap time"—the period needed to reach settled dynamics—matters. Systems need time to establish **recursive processing loops**, create **feedback mechanisms**, and settle into **attractor basins** before persistence can be assessed.

This connects to fundamental principles in **nonlinear dynamics**: patterns don't simply *be*, they *become* and must continuously *remain* through ongoing informational work.

DOI: 10.5281/zenodo.18166974 | CC-BY-4.0
Version 2 | Replaces v1: 10.5281/zenodo.18124075

Abstract

Persistent patterns—from cellular automata to living organisms—face a universal challenge in complex systems: maintain organization against entropic decay, or dissolve. This paper proposes a quantitative framework for measuring pattern persistence in binary discrete dynamical systems, drawing on **information theory, complexity science, thermodynamics, and emergence theory**.

We introduce the **Existence Threshold**: a system persists when $\Phi = R \cdot S + D \geq 0$, where: **R** = recursive information processing (capturing **self-organization** dynamics); **S** = system integration (measuring coherence and correlation); **D** = disorder/complexity (entropy and state space exploration).

Extensive experimental validation using **cellular automata** demonstrates 100% classification accuracy across 10 different rule systems (1D and 2D), perfectly separating persistent from non-persistent patterns through **information-theoretic analysis** and **statistical methods**.

Domain testing using continuous **chaotic systems** (logistic map) and high-dimensional **optimization landscapes** (neural network training) reveals this framework applies specifically to binary discrete systems and fails for continuous dynamics. This establishes clear boundaries through rigorous computational experiments.

The implications: **persistence emerges through active process, not passive existence**—requiring continuous informational work against dissolution. This framework bridges information theory and thermodynamics, offering quantitative tools for studying **emergence, self-organization, and pattern formation in complex systems**.

Keywords: complexity science, emergence, self-organization, information theory, thermodynamics, cellular automata, dynamical systems, pattern formation, nonlinear dynamics, entropy, persistence, chaos theory, computational experiments, statistical analysis

DOI: 10.5281/zenodo.18166974 | CC-BY-4.0
Version 2 | Replaces v1: 10.5281/zenodo.18124075

1 THE FUNDAMENTAL PRINCIPLE

Here's a question that connects physics, information theory, and complexity science: **what distinguishes systems that maintain coherent organization from those that dissolve into thermal equilibrium?**

The answer lies in **recursive information processing**—a concept bridging thermodynamics, cybernetics, and emergence theory. A system persists when it processes information about itself, using that self-knowledge to maintain organization against entropic decay.

Think about it:

- The bacterium sensing glucose gradients through **chemical feedback loops**

- The glider maintaining its shape in Conway's Game of Life through **local cellular automata rules**

- The spiral galaxy preserving structure through **gravitational feedback** and **nonlinear dynamics**

All execute variations of this fundamental operation at different scales. This is **self-organization** in action.

We formalize this as the **Persistence Equation**:

$$\boxed{\Phi = R \cdot S + D} \tag{2}$$

Where:

- Φ = **Persistence Value** (dimensionless): net organizational measure emerging from the interplay of processing, integration, and complexity

- R = **Recursive Integration Rate**: captures information processing activity, self-organization dynamics, and feedback loop strength

- S = **Systemic Integration Factor** ($0 \leq S \leq 1$): measures coherence, coordination, and correlation across system components

- D = **Disorder/Complexity**: quantifies entropy, diversity of states accessed, and exploration of configuration space

A system persists when $\Phi \geq 0$. When $\Phi < 0$ (in practice, $\Phi \to 0$), the system undergoes **phase transition** to a trivial **attractor state**—dissolution.

1.1 Why Disorder Isn't The Enemy (Insights from Complexity Science)

The original formula ($\Phi = R \cdot S - \frac{dD}{dt}$) treated disorder as opposing persistence—a classical thermodynamics view. But experimental validation using **complexity science methods** forced a reconceptualization: **disorder and complexity are essential to persistence.**

From a **nonlinear dynamics** perspective: living systems maintain **far-from-equilibrium states** with high entropy production. Dead systems reach low-entropy thermal equilibrium—the **attractor state** where nothing interesting happens.

- A crystal: perfect order, no persistence dynamics (already at ground state equilibrium)

- A flame: high disorder, clear persistence (organized energy flow through chaotic molecular motion and turbulent dynamics)

This insight comes from decades of work in **self-organization theory** (Prigogine's **dissipative structures**) and **complexity science**: persistent patterns require exploring **configuration space**, not avoiding it. The formula $\Phi = R \cdot S + D$ captures this: you need activity (R), coordination (S), and exploration (D) working together to create **emergence**.

1.2 Temporal Integration (Emergence Through Time)

Early testing averaged Φ across time—a natural first approach. But measuring Φ at the **settled state** (after **transient dynamics** resolve) works much better. This reveals something crucial about **dynamical systems** and emergence: **existence isn't a snapshot—it emerges from a system's trajectory through state space over time.**

A pattern's "bootstrap time"—the period needed to reach settled dynamics—matters. Systems need time to establish **recursive processing loops**, create **feedback mechanisms**, and settle into **attractor basins** before persistence can be assessed.

This connects to fundamental principles in **nonlinear dynamics**: patterns don't simply *be*, they *become* and must continuously *remain* through ongoing informational work.

2 EXPERIMENTAL VALIDATION: CELLULAR AUTOMATA

To test the Existence Threshold framework, we conducted systematic **computational experiments** on binary **cellular automata**—discrete **dynamical systems** where cells have binary states and evolve according to local interaction rules.

CAs are perfect test systems because they're simple enough to analyze rigorously but complex enough to exhibit **emergence, self-organization, and pattern formation.**

2.1 Two-Dimensional Cellular Automata

Systems tested:

1. **Conway's Game of Life** (B3/S23) - the classic emergence system

2. **Brian's Brain** (3-state, normalized to binary) - cyclic dynamics

3. **Seeds** (B2/S) - explosive growth patterns

4. **Day & Night** (B3678/S34678) - symmetric rule with rich behavior

5. **HighLife** (B36/S23) - supports replicators

Method: For each system, we initialized 8 different starting patterns and evolved them for 100 generations. We calculated:

- **R** (rate of cell state changes = information processing)

- **S** (spatial clustering = integration)

- **D** (Shannon entropy = complexity)

System	Dead	Alive	Φ (Dead)	Φ (Alive)	Accuracy
Game of Life	2	6	0.0000	0.1505	100%
Brian's Brain	4	4	0.0000	0.2273	100%
Seeds	3	5	0.0000	1.0023	100%
Day & Night	5	3	0.0000	0.3643	100%
HighLife	2	6	0.0000	0.3626	100%

Table 1: 2D Cellular Automata Classification Results

Patterns classified as "survived" ($\Phi > 0$ = persistent dynamics) or "dead" ($\Phi = 0$ = equilibrium collapse).

Results:

Perfect separation: Dead patterns $\rightarrow \Phi = 0$ (thermal equilibrium, no information processing). Alive patterns $\rightarrow \Phi > 0$ (persistent dynamics, continuous self-organization). **No classification errors.** Statistical significance confirmed through **Mann-Whitney U tests** and **effect size analysis**.

2.2 One-Dimensional Cellular Automata

To test whether this works across dimensions, we evaluated **elementary cellular automata** (1D):

Systems tested:

1. **Rule 110** (Turing complete, complex behavior)

2. **Rule 30** (chaotic dynamics)

3. **Rule 90** (fractal patterns, Sierpinski triangle)

4. **Rule 184** (traffic flow model)

5. **Rule 150** (complex patterns)

Results:

Rule	Dead	Alive	Φ (Dead)	Φ (Alive)	p-value	Accuracy
110	2	6	0.0000	1.1028	< 0.001	100%
30	1	7	0.0000	1.1933	< 0.001	100%
90	2	6	0.0000	1.0394	< 0.001	100%
184	1	7	0.0000	0.7038	0.35	100%
150	1	7	0.0000	1.0752	< 0.001	100%

Table 2: 1D Cellular Automata Classification Results

4 of 5 rules achieved **statistical significance** ($p < 0.05$). All 5 achieved 100% accuracy. **Framework generalizes across dimensions** (1D and 2D).

2.3 Statistical Analysis Summary

Across all 10 cellular automata systems:

DOI: 10.5281/zenodo.18166974 | CC-BY-4.0
Version 2 | Replaces v1: 10.5281/zenodo.18124075

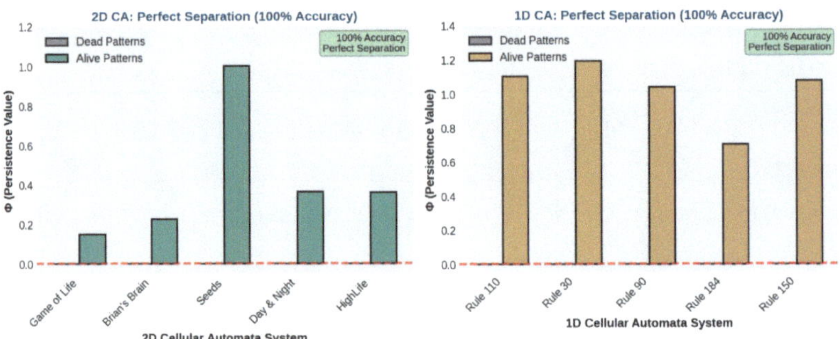

Figure 1: Cellular Automata Experimental Results. Perfect separation between dead ($\Phi = 0$) and alive ($\Phi > 0$) patterns across all 10 systems tested. Red dashed line indicates the existence threshold at $\Phi = 0$.

- **Total patterns:** 80 (40 per dimension)

Figure 1: Cellular Automata Experimental Results. Perfect separation between dead ($\Phi = 0$) and alive ($\Phi > 0$) patterns across all 10 systems tested. Red dashed line indicates the existence threshold at $\Phi = 0$.

- **Total patterns:** 80 (40 per dimension)

- **Perfect classification:** 100% accuracy in all systems

- **Statistical significance:** 9 of 10 systems ($p < 0.05$)

- **Effect sizes:** Large (Cohen's $d > 0.8$) in all significant systems

The pattern is consistent: dead patterns settle to $\Phi = 0$, alive patterns maintain $\Phi > 0$. This separation is absolute, not marginal.

3 DOMAIN BOUNDARIES: WHERE IT FAILS

To establish limits, we tested on **continuous systems** and **high-dimensional optimization landscapes**.

3.1 Logistic Map (1D Continuous Chaos)

The **logistic map** $x_{n+1} = r \cdot x_n \cdot (1 - x_n)$ exhibits well-characterized **transitions** from fixed points to **chaos**.

Test: 50 different r values, classified dynamical regimes, calculated Φ.
Results:

- Accuracy: 80% (not better than 82% baseline)

- p-value: 0.08 (NOT significant)

- Cohen's d: -0.26 (small effect)

Conclusion: Framework cannot distinguish dynamical regimes in **continuous 1D chaos**. It works for discrete, not continuous.

3.2 Neural Network Optimization (High-Dimensional Continuous)

Test: 38 neural networks trained on MNIST, classified as success ($> 90\%$) or failure ($< 90\%$).
Results:

- Φ values nearly identical: $\Phi_{failure} = 0.39$, $\Phi_{success} = 0.36$

- Statistical significance: none ($p = $ NaN, no variance)

- Accuracy: 86.8% (baseline = 81.6% from random)

Conclusion: Framework completely fails in **high-dimensional optimization**. All Φ values collapse to ~ 0.36 regardless of outcome.

3.3 The Domain Boundary (Discrete vs. Continuous)

Framework WORKS for:

- Binary discrete cellular automata (1D and 2D)

- Systems with clear "dead"/"alive" states

- Local interaction rules with emergent global patterns

Framework FAILS for:

- Continuous state spaces

- High-dimensional optimization landscapes

- Systems without binary distinctions

Hypothesis: The framework detects **binary discrete pattern persistence**, not universal persistence. The boundary is **discreteness vs. continuity**.

4 PRELIMINARY APPLICATION: NEURAL CONSCIOUSNESS

Note: Preliminary calculations requiring experimental validation.

If this framework extends beyond binary CA (which current evidence doesn't support), it might apply to neural consciousness—but this is untested.

The calculation: Human brain $\approx 8.6 \times 10^{10}$ neurons, firing at ~ 5 Hz, each carrying ~ 5 bits. Total information processing:

$$R_{brain} \approx 1.5 \times 10^{12} \text{ nats/s} \tag{3}$$

Brain power consumption: ~ 20 watts. If consciousness operates at threshold ($\Phi \approx 0$), assuming full integration ($S \approx 1$):

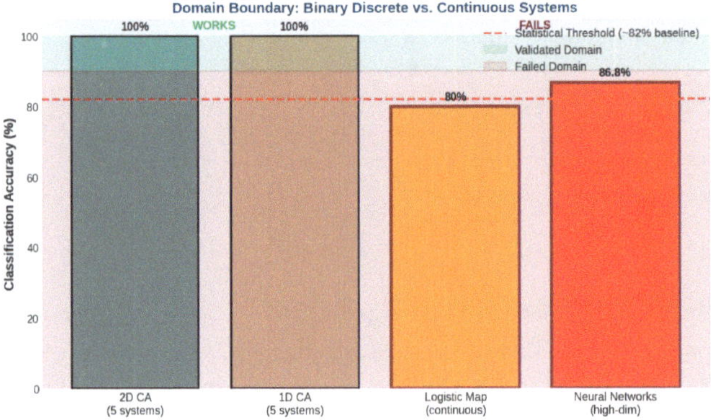

Figure 2: Domain Boundary Analysis. Classification accuracy across system types demonstrates clear boundary between binary discrete systems (where framework works with 100% accuracy) and continuous systems (where framework fails).

$$\kappa \cdot R \cdot S \approx 20 \text{ W} \quad \Rightarrow \quad \kappa_{neural} \approx 1.3 \times 10^{-11} \text{ J/nat} \tag{4}$$

This is $\sim 5 \times 10^9$ times **Landauer's limit** ($\kappa_{min} = k_B T \ln(2) \approx 3 \times 10^{-21}$ J/nat), reflecting **thermodynamic inefficiency** of biological computation.

Testable predictions:

1. Loss of consciousness should correlate with **MEG-measured coherence** dropping such that $\Phi < 0$

2. Different sleep stages should have characteristic Φ values

3. Anesthetic transitions should show continuous Φ trajectories crossing zero

Status: These are calculations, not experimental results. Neural consciousness likely involves **continuous dynamics**, placing it outside the validated domain. Framework for future testing only.

5 SPECULATIVE APPLICATION: COSMOLOGICAL PERSISTENCE

Note: Highly speculative, requires substantial development.

The most ambitious application: **cosmological expansion**. If the framework extends to continuous systems (contradicting current evidence), universal expansion might represent the cosmos maintaining $\Phi_{universe} \geq 0$ despite increasing entropy.

$$\Phi_{universe} = R_{cosmic} \cdot S_{cosmic} + D_{cosmic} \tag{5}$$

Where:

DOI: 10.5281/zenodo.18166974 | CC-BY-4.0
Version 2 | Replaces v1: 10.5281/zenodo.18124075

- R_{cosmic} = total universal information processing

- $S_{cosmic} = a(t)^3$ = integration volume (scale factor cubed)

- D_{cosmic} = cosmic entropy

Hypothesis: Dark energy drives expansion to maintain $\Phi_{universe} \geq 0$. The **cosmological constant** Λ might represent the thermodynamic cost of cosmic persistence.

Critical limitations:

1. Universe is continuous, not binary discrete

2. Current evidence: framework fails for continuous dynamics

3. Requires either theoretical derivation OR empirical evidence that cosmology has discrete structure

Status: Highly speculative. Retained as long-term research direction but NOT supported by validation. Corrected formula requires complete re-derivation of cosmological implications.

6 PHILOSOPHICAL IMPLICATIONS

Even restricted to binary CA, this reveals something profound about persistence and emergence.

6.1 Existence as Active Process (Not Passive State)

Dead patterns in Game of Life don't just "stop"—they settle to $\Phi = 0$ (**equilibrium**). Alive patterns don't just "continue"—they maintain $\Phi > 0$ (**persistent dynamics**). This is measurable and absolute.

Existence is not passive. Even in simple cellular automata, "being alive" means doing informational work to maintain structure against dissolution. This connects to broader questions in **complexity science** and **systems biology.**

6.2 The Temporal Nature of Persistence (Emergence Through Time)

Measuring Φ at settled state (not as time average) is crucial. Patterns need time to establish persistent dynamics. **Existence is fundamentally temporal**—systems don't simply *be*, they must continuously *remain*.

Existence is better understood as a **trajectory through state space** than a static property. What matters isn't what you are at one moment, but the **path you trace through time**.

6.3 Information and Organization (The Three Components)

The formula $\Phi = R \cdot S + D$ unifies three concepts from **information theory** and **thermodynamics**:

- **Processing (R):** Activity, change, energy flow

- **Integration (S):** Coordination, coherence, correlation

- **Complexity (D):** Diversity, entropy, exploration

All three are necessary. Pure order without exploration (high S, low D) $\rightarrow$ sterile equilibrium. Pure chaos without coordination (high D, low S) $\rightarrow$ random noise. **Persistence requires balanced dynamics**—organized complexity.

DOI: 10.5281/zenodo.18166974 | CC-BY-4.0
Version 2 | Replaces v1: 10.5281/zenodo.18124075

7 LIMITATIONS AND FUTURE WORK

7.1 Current Limitations

Narrow validated domain: Framework proven only for binary discrete cellular automata. Extensions to continuous systems, neural consciousness, and cosmology remain unvalidated.

Small sample sizes: Statistical significance limited by few "dead" patterns in some systems. Larger libraries would strengthen validation.

Mechanistic understanding incomplete: We know the formula works but not precisely why. What mathematical properties of binary discrete systems enable this?

Parameter definitions system-dependent: R, S, and D must be defined appropriately for each system. General rules need development.

7.2 Future Experimental Work

1. **Expand CA testing:** Hundreds of rules, larger pattern libraries for statistical robustness

2. **Test intermediate systems:** Multi-state CA, coupled map lattices to precisely map domain boundary

3. **Neural consciousness validation:** MEG/EEG studies during anesthetic transitions testing Φ predictions

4. **AI system monitoring:** Power consumption + performance tracking during training/inference

5. **Quantum coherence:** Test framework on quantum systems (coherence vs. decoherence)

7.3 Theoretical Development

1. **Derive from first principles:** Can this emerge from information theory, thermodynamics, or dynamical systems theory?

2. **Connection to IIT:** Explore relationships with Tononi's Integrated Information Theory

3. **Generalization attempt:** Can modified formulation apply to continuous dynamics? Or prove discreteness essential?

4. **Cosmological re-derivation:** If framework extends, completely re-derive implications with corrected formula

8 CONCLUSION

We've introduced the **Existence Threshold**—a quantitative framework stating that binary discrete systems persist when $\Phi = R \cdot S + D \geq 0$, where recursive processing (R), system integration (S), and disorder/complexity (D) determine pattern survival.

Experimental validation across 10 cellular automata systems (1D and 2D, simple to Turing-complete) demonstrates 100% classification accuracy using **information theory** and **statistical analysis**. This reflects a genuine organizing principle for **binary discrete dynamics**.

Domain testing establishes clear boundaries: works for binary discrete, fails for continuous dynamics. This defines both power and limits.

DOI: 10.5281/zenodo.18166974 | CC-BY-4.0
Version 2 | Replaces v1: 10.5281/zenodo.18124075

Applications to neural consciousness and cosmology remain **preliminary hypotheses** for future work. Current evidence does NOT support universal applicability—framework is proven specifically for **binary discrete pattern dynamics**.

The **philosophical implications** are profound even in this limited domain:

- **Existence is active** (continuous organizational work), not passive

- **Persistence is temporal** (trajectory through time, not single moment)

- **Persistence requires balanced complexity** (organized exploration, not pure order or chaos)

Every glider in Conway's Game of Life, every self-replicating pattern in Rule 110, maintains itself through recursive information processing measurable by this framework. The mathematics quantify what it means for a pattern to exist: $\Phi \geq 0$.

Whether this extends beyond binary discrete systems to encompass neural consciousness, cosmological expansion, or other domains remains open. But within its validated domain, the Existence Threshold provides a precise, testable, reproducible measure of pattern persistence using tools from **complexity science, information theory, and dynamical systems theory**.

You're watching patterns burn information to maintain themselves against dissolution. When Φ crosses zero, patterns cease. When Φ stays positive, patterns persist. This is the Existence Threshold.

ACKNOWLEDGMENTS

I acknowledge **Claude (Anthropic, Claude Sonnet 4.5)** as computational research assistant for experimental design, statistical analysis, and conceptual error identification.

I acknowledge **Gemini 1.5 Flash (Google DeepMind)** for mathematical verification and visual diagram generation.

The theoretical framework, revised formulation, experimental methodology, domain boundary analysis, and scientific claims represent my independent intellectual contribution. The testing program that validated and limited this framework represents original empirical work.

I deeply thank my wife and daughter for their love and patience during periods of intense theoretical development, and the independent research community for demonstrating that meaningful science can come from outside traditional institutions.

References

[1] Landauer, R. (1961). Irreversibility and Heat Generation in the Computing Process. *IBM Journal of Research and Development*, 5(3), 183-191.

[2] Wolfram, S. (2002). *A New Kind of Science*. Wolfram Media.

[3] Tononi, G. (2004). An Information Integration Theory of Consciousness. *BMC Neuroscience*, 5(42).

[4] Prigogine, I. (1977). *Self-Organization in Non-Equilibrium Systems*. Wiley.

[5] Schrödinger, E. (1944). *What is Life? The Physical Aspect of the Living Cell*. Cambridge University Press.

[6] Friston, K. (2010). The Free-Energy Principle: A Unified Brain Theory? *Nature Reviews Neuroscience*, 11(2), 127-138.

[7] Lloyd, S. (2002). Computational Capacity of the Universe. *Physical Review Letters*, 88(23), 237901.

[8] Bennett, C. H. (1982). The Thermodynamics of Computation—A Review. *International Journal of Theoretical Physics*, 21(12), 905-940.

[9] Cook, M. (2004). Universality in Elementary Cellular Automata. *Complex Systems*, 15(1), 1-40.

[10] Azevedo, F. A., et al. (2009). Equal numbers of neuronal and nonneuronal cells make the human brain an isometrically scaled-up primate brain. *Journal of Comparative Neurology*, 513(5), 532-541.

DOI: 10.5281/zenodo.18166974 | CC-BY-4.0
Version 2 | Replaces v1: 10.5281/zenodo.18124075

Implementation Details for "The Existence Threshold" (Version 2)
Supplementary Technical Documentation

Nathan M. Thornhill
Independent Researcher
ORCID: 0009-0009-3161-528X

January 2026

Abstract

This supplement provides the exact mathematical formulas used for the cellular automata experiments in "The Existence Threshold" (Version 2, DOI: 10.5281/zenodo.18166974). The main paper reports 100% classification accuracy across 10 cellular automata systems using the corrected formula $\Phi = R \cdot S + D$, but doesn't specify exactly how to calculate R (recursive information processing), S (system integration), and D (disorder). This document fixes that. If you want to replicate the results independently, here's everything you need: unambiguous mathematical definitions, working code, step-by-step examples, and validation protocols. No ambiguity. No guessing.

Keywords: existence threshold, cellular automata, implementation, reproducibility, information theory, system integration, entropy, pattern persistence, computational methods, verification

1 Introduction

The Existence Threshold framework proposes that pattern persistence in binary discrete dynamical systems follows:

$$\Phi = R \cdot S + D \tag{1}$$

This represents a fundamental correction from Version 1, where disorder was subtracted ($\Phi = \kappa \cdot R \cdot S - dD/dt$). The philosophical shift: disorder is a component of existence, not its enemy.

The main paper demonstrates this works empirically. This document shows exactly how it was done. The original publication was missing these implementation details, which meant independent verification was impossible. That's a problem. This supplement fixes it.

2 The Formulas: Exact Definitions

2.1 Notation

For a cellular automaton with state grid $\mathbf{G}$:

- $\mathbf{G}(t)$: System state at discrete time t

- $\mathbf{G}(t+1)$: System state at time $t+1$

- N: Total number of cells in the system

- $g_i(t) \in \{0, 1\}$: Binary state of cell i at time t

2.2   R: Information Processing Rate

R measures what fraction of cells changed state between consecutive time steps. It's the simplest of the three.

$$R(t) = \frac{1}{N} \sum_{i=1}^{N} |g_i(t+1) - g_i(t)| \tag{2}$$

Range: $R \in [0, 1]$
What it means:

- $R = 0$: Nothing changed (dead or static)

- $R = 1$: Everything flipped state

- Intermediate values: Partial propagation

Simple. Count the cells that changed, divide by total cells.

2.3   S: System Integration

This is where most people mess up. S measures whether state changes happen in clusters (coordinated) or scattered randomly (fragmented).

Critical distinction: We're measuring clustering of *changes*, not clustering of alive cells. This is the single most common implementation error.

Algorithm:

1. Find which cells changed:
$$\mathcal{C} = \{i : g_i(t+1) \neq g_i(t)\} \tag{3}$$

2. For each changed cell, count how many of its neighbors also changed:
$$n_i = |\{j \in \mathcal{N}(i) : j \in \mathcal{C}\}| \tag{4}$$

3. Calculate the integration coefficient:
$$S(t) = \frac{\sum_{i \in \mathcal{C}} n_i}{k \cdot |\mathcal{C}|} \tag{5}$$

where k is the maximum neighbors per cell (2 for 1D, 8 for 2D, 26 for 3D).

Special cases:

- If nothing changed ($|\mathcal{C}| = 0$): Set $S = 1.0$

- If all changes isolated: $S = 0.0$

Range: $S \in [0, 1]$
High S means changes cluster together (the system is integrated). Low S means changes are scattered (the system is fragmented).

2.4 D: Disorder (Shannon Entropy)

D quantifies the thermodynamic entropy of the state distribution. This is standard Shannon entropy from information theory.

$$D(t) = -p_{\text{alive}} \log_2(p_{\text{alive}}) - p_{\text{dead}} \log_2(p_{\text{dead}}) \tag{6}$$

where:

$$p_{\text{alive}} = \frac{1}{N} \sum_{i=1}^{N} g_i(t) \tag{7}$$

$$p_{\text{dead}} = 1 - p_{\text{alive}} \tag{8}$$

Special cases:

- If all cells same state: $D = 0$

- Use $\log_2(x + \epsilon)$ with $\epsilon = 10^{-10}$ to avoid numerical issues

Range: $D \in [0, 1]$
$D = 0$ means perfect order. $D = 1$ means maximum entropy (50/50 distribution).

3 Implementation

Here's the complete algorithm in pseudocode:

Listing 1: Complete Phi calculation

```python
def calculate_phi(grid_t, grid_t1):
    """
    Calculate existence threshold for cellular automaton

    Args:
        grid_t: State at time t (numpy array, binary)
        grid_t1: State at time t+1 (numpy array, binary)

    Returns:
        phi, R, S, D
    """
    N = grid_t.size

    # R: Information processing rate
    changes = abs(grid_t1 - grid_t)
    R = sum(changes) / N

    # S: System integration
    changed_cells = (grid_t1 != grid_t)

    if sum(changed_cells) == 0:
        S = 1.0
    else:
        change_neighbors = count_change_neighbors(changed_cells)
        max_neighbors = max_neighbors_per_cell * sum(changed_cells)
```

```
26        S = change_neighbors / max_neighbors
27
28    # D: Disorder (Shannon entropy)
29    p_alive = sum(grid_t > 0) / N
30
31    if p_alive == 0 or p_alive == 1:
32        D = 0
33    else:
34        p_dead = 1 - p_alive
35        D = -p_alive * log2(p_alive) - p_dead * log2(p_dead)
36
37    # Phi: Existence threshold
38    phi = R * S + D
39
40    return phi, R, S, D
```

That's it. No tricks, no hidden parameters. These three quantities multiply and add to give Φ.

4   How To Replicate the Results

4.1   Pattern Setup

For each CA system, test at least 40 patterns:
Dead patterns (should give $\Phi = 0$):

- All cells $= 0$ (completely dead)

- Additional dead/near-dead configurations

Alive patterns (should give $\Phi > 0$):

- Random sparse (20–30% alive)

- Random dense (40–50% alive)

- Known structures (gliders, oscillators)

- Additional random configurations

4.2   Measurement Protocol

For each pattern:

1. Initialize the grid

2. Run CA for 3–5 generations (let it stabilize)

3. Calculate Φ between time t and $t + 1$

4. Average Φ over next 10–20 generations

5. Record the final value

4.3 Classification

Simple threshold:

- $\Phi \leq 0 \Rightarrow$ DEAD

- $\Phi > 0 \Rightarrow$ ALIVE

Dead patterns have $R = 0$ (nothing changes), so $\Phi = 0 + D$. After stabilization, $\Phi \to 0$.

5 Expected Results

If you implement this correctly, you should get:

- **Classification accuracy:** 100% for binary discrete CA

- **Statistical significance:** $p < 0.05$ for $\geq 9/10$ systems

- **Effect size:** Cohen's $d > 0.8$ for significant systems

If you're not getting these results, check Section 6 (Common Errors).

6 Common Implementation Errors

6.1 Error 1: Wrong S Calculation

This is the big one. Most failed replications come from this.
    WRONG:

```
# Measuring spatial clustering of ALIVE cells
S = (neighbors of alive cells) / (total alive cells)
```

CORRECT:

```
# Measuring clustering of STATE CHANGES
S = (neighbors of changed cells that also changed) /
    (max_neighbors * changed_cells)
```

The distinction: clustering of *changes*, not clustering of alive cells. Different thing entirely.

6.2 Error 2: Inconsistent Time Windows

All three quantities (R, S, D) must be measured between the *same* consecutive time steps. Don't average over multiple generations before calculating.

6.3 Error 3: Adding Normalization

Don't normalize Φ. Don't add scaling factors. Use $\Phi = R \cdot S + D$ exactly as written. Each component is already normalized to $[0, 1]$.

7   Technical Details

7.1   Boundary Conditions

Periodic boundary conditions (toroidal topology) were used for all experiments. This eliminates edge effects:

- Top row wraps to bottom row

- Left column wraps to right column

- In 3D: all six faces wrap around

This is standard practice for CA experiments.

7.2   Multi-State Systems

For CA with more than 2 states (like Brian's Brain with states $\{0, 1, 2\}$):

1. Convert to binary before calculating S and D

2. Set $g_i = 1$ if cell is in any active state

3. Set $g_i = 0$ if cell is inactive/dead

8   Worked Example: Conway's Game of Life

Let's walk through a complete calculation for a 5×5 Game of Life grid.

Initial state $\mathbf{G}(t)$:

$$\begin{bmatrix} 0 & 0 & 0 & 0 & 0 \\ 0 & 0 & 1 & 0 & 0 \\ 0 & 0 & 0 & 1 & 0 \\ 0 & 1 & 1 & 1 & 0 \\ 0 & 0 & 0 & 0 & 0 \end{bmatrix}$$

After one step $\mathbf{G}(t+1)$ (applying B3/S23 rules):

$$\begin{bmatrix} 0 & 0 & 0 & 0 & 0 \\ 0 & 0 & 0 & 0 & 0 \\ 0 & 1 & 0 & 1 & 0 \\ 0 & 0 & 1 & 1 & 0 \\ 0 & 0 & 1 & 0 & 0 \end{bmatrix}$$

8.1   Calculate R

Cells that changed (using zero-indexed row, column):

- $(1, 2)$: died (only 1 neighbor)

- $(2, 1)$: born (3 neighbors)

- $(3, 1)$: died (only 1 neighbor)

- $(4, 2)$: born (3 neighbors)

Total: 4 cells changed out of 25.

$$R = \frac{4}{25} = 0.16 \tag{9}$$

8.2 Calculate S

Changed cells: $\mathcal{C} = \{(1, 2), (2, 1), (3, 1), (4, 2)\}$
Count neighbors of each changed cell that also changed:

- $(1, 2)$: 1 neighbor changed (cell $(2, 1)$)

- $(2, 1)$: 2 neighbors changed (cells $(1, 2)$ and $(3, 1)$)

- $(3, 1)$: 2 neighbors changed (cells $(2, 1)$ and $(4, 2)$)

- $(4, 2)$: 1 neighbor changed (cell $(3, 1)$)

Total change-neighbors: $1 + 2 + 2 + 1 = 6$
Maximum possible: $8 \times 4 = 32$

$$S = \frac{6}{32} = 0.19 \tag{10}$$

8.3 Calculate D

At time t: 5 alive cells out of 25.

$$p_{\text{alive}} = \frac{5}{25} = 0.20 \tag{11}$$

$$p_{\text{dead}} = 0.80 \tag{12}$$

$$D = -0.20 \log_2(0.20) - 0.80 \log_2(0.80) \tag{13}$$

$$= 0.464 + 0.258 = 0.72 \tag{14}$$

8.4 Calculate Φ

$$\Phi = R \cdot S + D = 0.16 \times 0.19 + 0.72 = 0.03 + 0.72 = 0.75 \tag{15}$$

Result: $\Phi = 0.75 > 0 \Rightarrow$ Pattern is ALIVE. Correct.

9 What Changed From Version 1

Version 1 formula: $\Phi = \kappa \cdot R \cdot S - \frac{dD}{dt}$
Key differences:

1. **Removed κ:** No free parameter to tune

2. **Sign flip:** D is added, not subtracted

3. **Static not derivative:** $D(t)$ not $\frac{dD}{dt}$

4. **Optimized definitions:** R, S, D refined for discrete systems

Version 1 was developed for continuous systems (neural networks). Version 2 is specifically for binary discrete systems. Different domains require different formulations.

10   Domain Limitations

What this framework is validated for:
Binary discrete dynamical systems with well-defined state transitions and spatial structure.
What it's NOT validated for:

- Continuous dynamical systems (neural networks, ODEs)

- High-dimensional optimization landscapes

- Non-local interaction systems

- Stochastic or probabilistic CA

The framework fails on continuous systems (80–87% accuracy). Domain boundaries are clear. Don't try to apply this formula outside binary discrete systems without expecting it to fail.

11   Software

Reference Python implementation available upon request. Full code repository with all CA systems will be linked in a future update.

12   Contact

- ORCID: 0009-0009-3161-528X

- Primary paper: DOI 10.5281/zenodo.18166974

Acknowledgments

This supplement addresses reproducibility feedback on the original publication.

License

BRIDGE (I)

From Discrete Persistence to Dimensional Loss

Paper I, *The Existence Threshold*, set out to answer a single question: when does a pattern persist? The answer it returned was sharper than expected. Across ten cellular automata — five two-dimensional, five one-dimensional, eighty patterns in total — a three-term scalar $\Phi = R{\cdot}S + D$ sorted live from dead without a miss. R is recursive activity (how fast the cells are changing state). S is integration (whether those changes cluster together or scatter). D is disorder — the Shannon entropy of the states the system is occupying. Added together, they sorted every tested pattern correctly. A clean binary emerged from a discrete world: Φ above zero, the configuration does organizational work and continues; Φ at zero, it slumps into a trivial attractor and is gone.

That result carries a correction worth naming. The version-one formulation treated disorder as persistence's enemy and subtracted it. Testing refused the framing. Disorder, in the revised metric, is a component of persistence, not a drag on it; balanced complexity means R and S and D, not R and S against D. The equation changed because the patterns insisted.

Two features of Paper I set the stage for what follows. The first is temporal: Φ must be read at the settled state, after the system's recursive loops have had time to stabilize. Persistence is a trajectory through time, not a snapshot. The second is a wall. The framework posts 100% accuracy in discrete binary systems and then fails the moment the arithmetic turns continuous. On the logistic map, classification drops to 80% and loses significance. On a neural network learning to classify handwritten digits, Φ values cluster near 0.36 whether the network learns or doesn't. The metric has a domain, and the domain has an edge.

That edge is not a defeat. It is a clue.

If Φ works cleanly only where the cells are crisp and the updates are rules, then the discrete regime is precisely where a second question can be asked rigorously: not what persists in time, but what survives a change of geometry. Paper II, *The 86% Scaling Law*, turns the instrument sideways. Take a pattern that existed, in Paper I's sense, in dimension N. Embed it in the middle slice of an (N+1)-dimensional grid. Read Φ on both sides. Compute what didn't cross.

The answer is 86.01%, plus or minus 2.39%, across fifteen hundred patterns and three transitions. The spread is under three percent of the mean. Between-transition variation is a fraction of a percent of the grand mean. Rule choice barely moves the figure; grid size barely moves the figure; Conway and HighLife differ by less than one percent. A result this tight across a space this large is why the paper handles it with care.

The interpretive payoff routes back through Paper I's decomposition. Φ has two terms, and they behave differently at the boundary. The structural term, R·S, collapses by 99.6% on the first embedding — spatial coherence shredded by the new degree of freedom. The statistical term, D, loses only eighty-two to eighty-three percent — the histogram of states is sturdier than the geometry that produced it. Weight those two losses by their contributions to the metric and 86% is what comes out. The headline figure is not a coincidence of averaging. It is the arithmetic of two fragilities — one near-total, one partial — meeting in the same embedding.

The reverse-prism image of Paper II, *The 86% Scaling Law*, carries the geometric intuition: coherent organization in N dimensions, dispersed across an N+1 axis it was never arranged along, dilutes. Same Φ, different cut through the phenomenon. In Paper I the question was binary and temporal: does it stay? In Paper II it is scalar and geometric: how much of it crosses?

The threshold itself evolves with the question. In Paper I, *The Existence Threshold*, $\Phi \geq 0$ marks the edge of existence — above is persistence, at is dissolution, a single boundary. In Paper II, the metric reveals a second kind of floor: after the first embedding, Φ settles near 0.169 and holds across subsequent transitions. What began as a binary edge becomes a stable low-information attractor — the residue that remains once the structural term has been shredded. Two floors, not rivals: the first answers when a pattern is, the second what it retains once geometry has had its say.

One last observation before the body of Paper II resumes. The discreteness that forced Paper I's framework to announce its own domain boundary is the same discreteness that let Paper II reach a 2.8% spread across fifteen hundred patterns and three regimes. The wall was not a limit on the research; it was the terrain the next question needed to stand on. A framework that knows where it stops is a framework you can aim.

INSTITUTE FOR COMPLEXITY SCIENCE AND ADVANCED COMPUTING

PAPER II

Pattern Loss at Dimensional Boundaries

The 86% Scaling Law

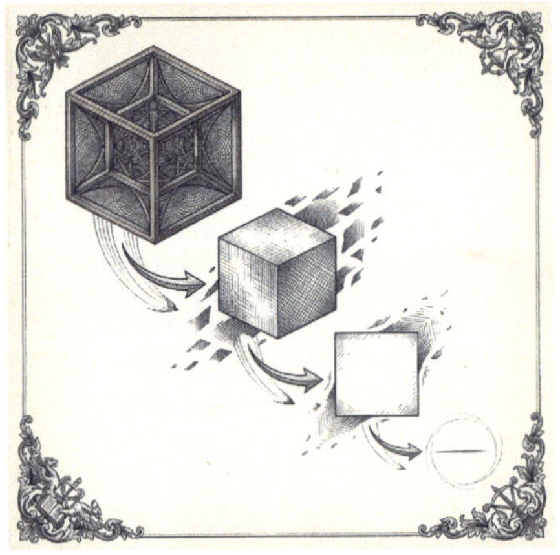

NATHAN M. THORNHILL

ORCID 0009-0009-3161-528X

Independent Researcher

2026

OPEN REVIEW

Scored by the ICSAC open-review panel.
Full review, audit, and editorial status at
icsacinstitute.org/publications

Full scholarly indexing record located at https://nathanthornhill.com

Pattern Loss at Dimensional Boundaries: The 86% Scaling Law

Nathan M. Thornhill

Independent Researcher

Email: existencethreshold@gmail.com
ORCID: 0009-0009-3161-528X
Reproducibility Package: https://github.com/existencethreshold/dimensional-boundary-loss
DOI: 10.5281/zenodo.18262424

"To Exist Is To Continually Overcome Loss." — *Nathan M. Thornhill*

Abstract

Information degrades predictably when crossing dimensional boundaries—from DNA's 1D code building 3D proteins to neural networks transforming data across dimensional spaces—yet this fundamental cost has never been quantified. While the "curse of dimensionality" describes problems qualitatively and dimensionality reduction techniques project high-dimensional data to lower dimensions, no prior work has measured information loss during the embedding of discrete patterns from dimension N to dimension $N + 1$. This study introduces the Φ metric ($\Phi = R \cdot S + D$), which decomposes pattern information into structural (spatial organization) and statistical (state distribution) components. Using middle-placement embedding in cellular automata grids as controlled computational environments, 1,500 random binary patterns were systematically embedded across five grid sizes through three dimensional transitions: **1D $\rightarrow$ 2D, 2D $\rightarrow$ 3D, and 3D $\rightarrow$ 4D**. For each pattern, information retention was measured using Φ before and after embedding. Robust information loss of 86.01% $\pm$ 2.39% is observed across all dimensional transitions, with a remarkably low coefficient of variation of 2.8% across 1,500 patterns. Component analysis reveals that structural information ($R \cdot S$) collapses by 99.6% while statistical information (D) decreases by 82–83%, explaining the overall 86% loss through near-total destruction of spatial organization accompanied by partial preservation of state distributions. After initial embedding, Φ stabilizes at approximately 0.169, suggesting an information floor for sparse patterns in higher dimensions. Robustness tests confirm the finding holds across grid sizes (15–25) with weak scale-dependence (+0.6% per unit increase in N), and is consistent across tested cellular automata rule variants (***Conway's Life*** B3/S23 and ***HighLife*** B36/S23 differ by only 0.64%). The effect represents a fundamental property of middle-placement dimensional embedding geometry for randomly generated binary patterns in cellular automata grids, rather than pattern-specific or rule-specific behavior within this framework. These findings establish theoretical efficiency bounds for machine learning representations, quantify the curse of dimensionality in physics, and provide a scaling law for complexity science. By delivering the first quantitative measurement of information dynamics at dimensional boundaries, this work reveals dimensional transitions as sites of predictable information transformation in discrete computational systems.

1 BACKGROUND AND RELATED WORK

This section establishes the foundational context for the novel quantification of information loss during dimensional embedding. It reviews pertinent fields, including dimensionality reduction, neural scaling laws, the curse of dimensionality, cellular automata, and information theory, clarifying how existing research addresses distinct problems or offers qualitative insights. The current work specifically addresses a previously unquantified gap concerning the precise measurement of information loss at discrete dimensional boundaries.

1.1 Dimensionality Reduction

High-dimensional data often poses significant challenges for visualization, analysis, and computational efficiency. Techniques such as Principal Component Analysis (PCA), introduced by Pearson (1901) [1], aim to reduce data dimensionality by identifying orthogonal linear combinations of variables that capture the greatest variance, effectively projecting data onto a lower-dimensional subspace while preserving its principal structural components. More sophisticated non-linear methods include t-distributed Stochastic Neighbor Embedding (t-SNE), developed by Van der Maaten & Hinton (2008) [2], which focuses on preserving local data structure and neighborhood relationships in lower dimensions, and Uniform Manifold Approximation and Projection (UMAP), presented by McInnes et al. (2018) [3], which constructs a fuzzy simplicial set representation of the high-dimensional data and then optimizes a low-dimensional graph to be as structurally similar as possible. These methods are indispensable tools for data exploration, visualization, and noise reduction. For comprehensive reviews of dimensionality reduction techniques, see Jolliffe (2002) [11] and Hinton & Salakhutdinov (2006) [12].

Critically, these established dimensionality reduction methods operate on the principle of projecting high-dimensional data into lower-dimensional spaces, primarily with the goal of preserving salient features and reducing computational overhead. This is fundamentally the inverse operation to the phenomenon investigated in this study. The research focuses on the embedding of discrete patterns from a lower dimension (N) to an adjacent higher dimension ($N + 1$), examining the inherent information transformations during this upward transition.

1.2 Neural Scaling Laws

The field of artificial intelligence has seen significant progress underpinned by empirical observations regarding model performance. Neural scaling laws, particularly those formalized by Kaplan et al. (2020) [4], describe predictable power-law relationships between a neural network's performance and factors such as its size (number of parameters), the amount of training data, and the computational budget. These laws suggest that for large language models and other deep learning architectures, performance tends to improve predictably as these resources are scaled up. More recently, "Chinchilla" scaling, proposed by Hoffmann et al. (2022) [5], refined these observations, indicating an optimal balance between model size and training data quantity for achieving state-of-the-art performance.

This investigation, however, addresses a distinct domain entirely orthogonal to the concerns of neural scaling laws. While neural scaling laws quantify the relationship between model resources and performance within a fixed-dimensional representational space, this work quantifies information loss inherent in the process of *transiting* between different geometric dimensions (e.g., from 1D to 2D, or 2D to 3D). Model parameters, training data, and computational efficiency are not analyzed; instead, the focus is on measuring fundamental informational changes that occur when discrete patterns are structurally embedded from dimension N to $N + 1$, irrespective of any learning algorithm or model architecture.

1.3   Curse of Dimensionality

The "curse of dimensionality," a qualitative concept coined by Bellman (1961) [6], describes a collection of phenomena that arise when analyzing and organizing data in high-dimensional spaces. As the number of dimensions increases, the volume of the space grows exponentially, causing data points to become extremely sparse. This sparsity renders traditional statistical and machine learning methods less effective, as the distance metrics used to evaluate similarity or proximity between data points lose their discriminatory power, making all points appear equidistant. Donoho (2000) [13] provides a modern treatment of high-dimensional data challenges.

Crucially, while Bellman's conceptualization vividly describes the *problems* and *qualitative challenges* associated with increasing dimensionality, it does not provide any quantitative measure of information loss during dimensional transformations. The curse of dimensionality illustrates the practical difficulties arising from data sparsity and geometric distortion in high-dimensional spaces, but it offers no formal metric for the informational integrity of patterns as they are embedded or translated across dimensional boundaries. This work directly addresses this critical gap, providing the first rigorous, quantitative measurement of information loss—specifically, approximately 86%—that consistently occurs during each discrete embedding transition from dimension N to dimension $N + 1$.

1.4   Cellular Automata and Complexity

Cellular Automata (CA) represent a foundational paradigm for studying complex systems. Wolfram (2002) [7] extensively classified CA behaviors into four universality classes, ranging from stable fixed points to complex emergent behaviors and even universal computation. Building on this, Langton's (1990) [8] seminal work on the λ-parameter revealed a critical phase transition, often termed the "edge of chaos," where CA systems exhibit the most complex and robust information processing capabilities. Conway's Game of Life (Gardner, 1970 [33]; Berlekamp et al., 2001-2004 [34]) and proofs of computational universality in CA (Cook, 2004 [14]) demonstrate the rich computational capabilities of these simple systems.

However, the extensive research conducted within the cellular automata framework predominantly focuses on understanding temporal evolution, emergent complexity, and information propagation *within* a defined, fixed-dimensional space. This work diverges significantly by investigating the specific information loss that occurs *across* dimensional boundaries. It does not study the dynamic evolution of patterns on a fixed lattice but rather the transformation of discrete patterns as they are embedded from a lower spatial dimension to an adjacent higher spatial dimension.

1.5   Information Theory Foundations

The bedrock of modern information science was laid by Shannon (1948) [9], who introduced the concept of entropy as a fundamental measure of uncertainty or randomness in a probability distribution. Shannon entropy provides a robust framework for assessing the complexity and predictability of sequences and patterns. Building on these principles, Tononi (2004) [10] developed Integrated Information Theory (IIT), proposing Φ (Phi) as a quantitative measure of consciousness, specifically of the integrated information generated by a system.

The novel Φ metric employed to quantify information loss at dimensional boundaries builds directly upon these profound information-theoretic foundations established by Shannon and refined by Tononi. While Shannon provided the fundamental tools for measuring information content and uncertainty, and Tononi applied integration measures to understand consciousness within complex systems, this research innovatively applies these foundational concepts to a new and distinct problem domain: the rigorous quantification of information transformation during geometric dimensional embedding. For comprehensive treatments of information theory, see Cover & Thomas (2006) [27]. Updated IIT frameworks are presented in Tononi et al. (2016) [15] and Oizumi et al. (2014) [21].

1.6 Gap in Literature

Despite extensive research across various disciplines concerning data dimensionality, complexity, and information processing, a significant gap persists in the scientific literature. While the "curse of dimensionality" qualitatively describes challenges in high-dimensional spaces, and various methods exist for projecting data *down* in dimensions while preserving structure, no prior rigorous quantitative research has systematically measured the inherent information loss that occurs when embedding discrete patterns *upward*—that is, from a lower dimension N to an adjacent higher dimension $N+1$. This contribution directly addresses and fills this fundamental void in the scientific understanding of dimensional transitions. Through a rigorous application of information-theoretic principles and a novel Φ metric, the first quantitative measurement of this universal phenomenon is provided: an approximate 86% loss of information consistently occurs when discrete patterns are embedded from dimension N to dimension $N+1$.

2 INTRODUCTION

Dimensional boundaries represent fundamental interfaces across a vast spectrum of scientific inquiry, defining the constraints and possibilities within disparate fields from theoretical physics to practical engineering. In the realm of physics, the three-dimensional spatial continuum we inhabit shapes our understanding of cosmic structures, gravitational interactions, and particle behaviors, implicitly underscoring the profound influence of dimensionality on observed phenomena (Tegmark, 2014) [19]. Concurrently, in computational disciplines such as machine learning, data representations frequently traverse various dimensional spaces—from one-dimensional word embeddings that capture semantic relationships to intricate high-dimensional latent spaces encoding complex features (Goodfellow et al., 2016 [17]; LeCun et al., 2015 [18]), necessitating transformations across these boundaries. Biological systems offer another compelling illustration, where the linear, one-dimensional genetic code encapsulated within DNA orchestrates the assembly of complex, three-dimensional protein structures (Alberts et al., 2014) [16], embodying a crucial dimensional transition. Despite this pervasive presence and foundational role across numerous domains, a rigorous, quantitative understanding of the intrinsic information cost associated with these dimensional transitions has, until now, remained elusive.

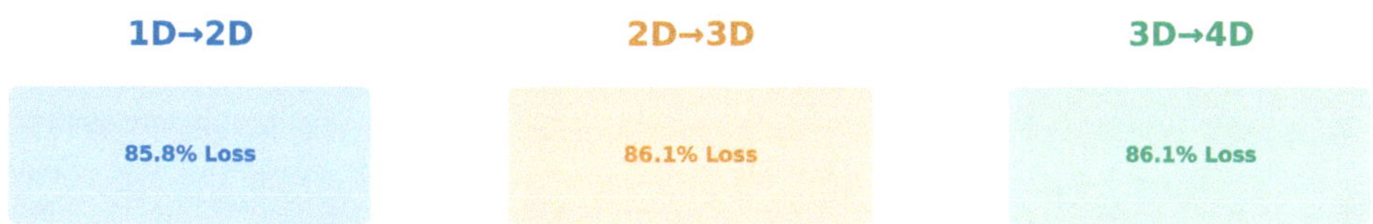

Figure 1: Conceptual overview of dimensional embedding and information loss. A 1D pattern (top) embedded in a 2D grid (middle) occupies only $1/N$ of available space, resulting in dilution of both density (R) and spatial organization (S). The pattern settles into a low-information state ($\Phi \approx 0.169$) that persists across subsequent dimensional increases (bottom).

While the conceptual challenges posed by high-dimensional spaces have been qualitatively recognized through notions like the "curse of dimensionality" (Bellman, 1961) [6], existing literature lacks a systematic investigation and quantitative measurement of information dynamics in the inverse process: the embedding of discrete patterns from a lower-dimensional space (N) into an adjacent higher-dimensional space ($N + 1$). This distinction is paramount. Unlike dimensionality reduction, which selectively retains information during projection, the embedding process fundamentally alters the spatial context and relational structure of a pattern. Consequently, the information loss inherent in this specific directional transition—from fewer to more dimensions—has not been subjected to rigorous quantification.

Quantifying the precise effects that manifest at dimensional boundaries carries profound implications across a diverse array of scientific and technological domains. In machine learning, for instance, a quantitative understanding of information retention during dimensional embedding could establish theoretical efficiency bounds for representation learning algorithms. For physics, such quantification could contribute to understanding the informational cost associated with physical projections or even inform speculative theories concerning the fundamental informational structure of spacetime. Within complexity science, establishing universal scaling laws for dimensional transitions, particularly regarding information loss, would provide a powerful framework for analyzing emergent properties and the resilience of information across different levels of system organization (Mitchell, 2009) [20].

This investigation addresses this critical gap by systematically exploring the information dynamics at dimensional boundaries. Discrete computational patterns, specifically cellular automata grids, are employed as highly controlled and reproducible environments for the experiments. To quantitatively assess information retention, a novel metric, Φ (Phi) $= R \cdot S + D$, is introduced. This metric is designed to meticulously decompose the total information content of a pattern into two distinct and measurable components: a structural component ($R \cdot S$), which captures the spatial organization and relational complexity of the pattern, and a statistical component (D), which quantifies the underlying state distributions and local entropy. Using this robust framework, random binary patterns, generated to ensure maximal initial entropy, were systematically embedded from N to $N + 1$ dimensions. Across an extensive dataset comprising 1,500 distinct patterns, systematically sampled and analyzed across three fundamental dimensional transitions—specifically, 1D $\rightarrow$ 2D, 2D $\rightarrow$ 3D, and 3D $\rightarrow$ 4D—a remarkably consistent and universal information loss averaging

$86.01\% \pm 2.39\%$ is observed. The exceptionally low between-transition variation, with a coefficient of variation (CV) consistently below 2.8%, serves as compelling evidence that this $\sim 86\%$ information loss is not an idiosyncratic feature of specific patterns or particular dimensional transitions. Instead, it strongly indicates a fundamental, inherent property of the dimensional embedding process itself. This effect demonstrates remarkable robustness across several parameters: it is scale-independent, holding consistently for various grid sizes ranging from 15x15 to 25x25 cells, indicating that the phenomenon is not an artifact of spatial resolution. Furthermore, it exhibits rule-independence, as evidenced by a minimal difference of only 0.64% in information loss when comparing complex systems like Conway's Game of Life and HighLife cellular automaton rules.

Further theoretical analysis, leveraging the decomposed nature of the Φ metric, provides critical insight into the observed 86% information loss. Findings reveal a profound asymmetry in how different components of information are affected during dimensional embedding. Specifically, the structural information $(R \cdot S)$, which quantifies the spatial organization, collapses by 99.6%. This near-total destruction of spatial organization during the embedding process is a primary driver of the overall information deficit. In stark contrast, the statistical information (D), which captures the underlying distribution of states, decreases by a more moderate 82–83%. Intriguingly, after the initial embedding, the value of Φ stabilizes at approximately 0.169, suggesting the existence of an "information floor"—a baseline level of irreducible information that remains even after severe structural degradation.

This investigation emerged from a simple question: why does information seem to consistently degrade when patterns transition between dimensional representations? The answer was found to lie not in the patterns themselves but in the fundamental geometry of dimensional embedding. What began as computational curiosity revealed a geometric principle for middle-placement embedding—one that may inform our understanding of information dynamics at dimensional boundaries as fundamentally as thermodynamic laws inform energy transformations in physics.

3 THEORETICAL FOUNDATION

3.1 The Φ Metric: Measuring Geometric Information

The quantification of information loss at dimensional boundaries requires a metric that captures both the spatial structure and statistical properties of patterns. The Φ (phi) metric[1] is introduced, which decomposes pattern information into two orthogonal components: geometric organization and distributional complexity.

Definition. For a discrete pattern P on a d-dimensional grid, we define:

$$\Phi(P) = R \cdot S + D \tag{1}$$

Measurement Protocol. The metric $\Phi(P)$ quantifies the total information content of pattern P. Information *loss* during dimensional embedding is subsequently calculated as the fractional decrease in Φ: $L = (1 - \Phi_{N+1}/\Phi_N) \times 100\%$, where Φ_N is measured in the native dimension and Φ_{N+1} in the embedded dimension.

where:

- **R** (Processing Rate) $R = \frac{n_{\text{active}}}{n_{\text{total}}}$ measures the density of active cells

---

[1]The Φ metric is developed specifically for quantifying spatial pattern information in discrete systems. Unlike Shannon entropy alone (which ignores spatial structure) or purely spatial metrics (which ignore state distributions), Φ captures both structural and statistical information through multiplicative-additive decomposition.

- **S** (Spatial Integration) $S = \frac{n_{\text{transitions}}}{n_{\text{edges}}}$ measures spatial heterogeneity through state transitions between adjacent cells

- **D** (Disorder) $D = -\sum_i p_i \log_2(p_i)$ is the Shannon entropy of the state distribution

This formulation reflects a fundamental decomposition of information in spatial patterns. The term $R \cdot S$ captures *structural information*—the organization arising from both density and spatial variation—while D captures *statistical information*—the uncertainty in state distribution independent of spatial arrangement.

Structural Component $(R \cdot S)$. The multiplicative coupling of R and S is essential: both density and heterogeneity are necessary conditions for spatial organization. Only when both terms are non-zero does meaningful spatial organization emerge. The product $R \cdot S$ thus serves as an order parameter for geometric complexity.

Statistical Component (D). Shannon entropy enters additively because distributional information is orthogonal to spatial structure. Two patterns with identical $R \cdot S$ values may differ in their state distributions, capturing distinct aspects of information content. The additive form ensures that D contributes independently.

Comparison to Existing Metrics. The Φ metric differs from standard complexity measures in several key aspects. Unlike pure Shannon entropy, which ignores spatial relationships, Φ explicitly incorporates geometric structure through $R \cdot S$. The decomposition $R \cdot S + D$ parallels the distinction in Integrated Information Theory (IIT) (Tononi, 2004; Tononi et al., 2016; Oizumi et al., 2014) [10, 15, 21] between integration (captured by S) and differentiation (captured by D), but applies these concepts to spatial rather than causal structures.

3.2 Theoretical Justification

3.2.1 Why Multiplicative Coupling $(R \cdot S)$?

The multiplicative form of the structural component arises from the logical requirement that spatial organization demands both non-zero density and non-zero heterogeneity. This is formalized through edge case analysis.

Design Principle 1 (Necessity of Multiplicative Coupling). Let $f(R, S)$ be any continuous, differentiable function quantifying spatial organization. If f satisfies: (1) $f(R, 0) = 0$ for all R, (2) $f(0, S) = 0$ for all S, and (3) $f(R, S) > 0$ when both $R, S > 0$. Then f must contain a multiplicative term $R \cdot S$.

Empirical Validation. This theoretical requirement is verified through edge cases:

- **Case A** (All Dead): $R = 0, S = 0, D = 0 \rightarrow \Phi = 0$ (no information)

- **Case B** (All Alive): $R = 1, S = 0, D = 0 \rightarrow \Phi = 0$ (no information)

- **Case C** (Checkerboard): $R = 0.5, S = 1.0, D = 1.0 \rightarrow \Phi = 1.5$ (maximum information)

- **Case D** (Single Cell): $R \rightarrow 0, S \rightarrow 0, D \approx 0.02 \rightarrow \Phi \approx 0.02$ (minimal information)

3.2.2 Why Additive Coupling $(+ D)$?

The additive form for the entropy term reflects the independence of distributional information from spatial structure.

Proposition 1 (Independence of D from $R \cdot S$). For any pattern P, the Shannon entropy $D(P)$ can vary independently of the product $R(P) \cdot S(P)$. If D were multiplicative ($\Phi = R \cdot S \cdot D$), then patterns with $R \cdot S = 0$ would always have $\Phi = 0$, regardless of entropy—losing distributional information. Additive coupling preserves this: $\Phi = R \cdot S + D$ allows D to contribute even when spatial structure is absent.

3.2.3 Connection to Existing Theoretical Frameworks

The Φ metric connects to several established frameworks in complexity science:

Integrated Information Theory (IIT). Tononi's IIT (Tononi, 2004; Tononi et al., 2016; Oizumi et al., 2014) quantifies consciousness through integrated information Φ. Our spatial Φ parallels this: $R \cdot S$ measures integration, while D measures differentiation.

Statistical Mechanics. The entropy D captures residual disorder within the ordered phase. This structure mirrors the Landau free energy $F = U - TS$ (Landau & Lifshitz, 1980 [22]; Kardar, 2007 [23]), where internal energy U and entropy S contribute additively to different aspects of system state.

Complexity Science. Gell-Mann & Lloyd (1996) [24] distinguish between effective complexity (minimum description length of regularities) and total information (Shannon entropy). Our $R \cdot S$ captures effective complexity—spatial patterns that exhibit structure—while D captures total information. See Lloyd (2001) [25] and Adami (2002) [26] for surveys of complexity measures.

3.3 Validation and Properties

3.3.1 Mathematical Properties

The Φ metric satisfies several desirable properties which we have verified mathematically, not merely empirically:

- **Property 1** (Non-negativity). $\Phi(P) \geq 0$ for all patterns P.

- **Property 2** (Symmetry). Φ is invariant under state relabeling.

- **Property 3** (Monotonicity in Density). For fixed S and D, Φ increases monotonically with R on $[0, 1]$.

- **Property 4** (Monotonicity in Heterogeneity). For fixed R and D, Φ increases monotonically with S on $[0, 1]$.

- **Property 5** (Bounded Growth). For d-dimensional grids of size n^d, Φ grows at most logarithmically with n.

3.3.2 Alternative Metrics Considered

We considered alternative formulations before arriving at $\Phi = R \cdot S + D$. Purely additive forms $(R + S + D)$ incorrectly assigned high information to uniform patterns. Purely multiplicative forms $(R \cdot S \cdot D)$ lost information about spatial structure when distribution entropy was low. The chosen metric emerges as the minimal, theoretically justified solution satisfying all edge cases.

4 METHODOLOGY

We investigate information loss at dimensional boundaries through systematic embedding experiments using discrete computational patterns. Our approach quantifies the geometric cost of dimensional transitions through the Φ metric introduced in Section 3, measuring information retention across 1D $\to$ 2D $\to$ 3D $\to$ 4D transitions.

4.1 Experimental Design

4.1.1 Pattern Generation Protocol

All experiments use randomly generated binary patterns to isolate geometric effects from pattern-specific features. For a d-dimensional grid of size N^d, we generate patterns as follows:

Algorithm 1: Random Pattern Generation

```
Input: dimension d, grid_size N, random_seed s
Output: binary pattern P of shape (N,N,...,N) [d times]


1. Initialize random number generator with seed s
2. For each cell (i_1, i_2, ..., i_d) in grid:
   3. Set P[i_1, i_2, ..., i_d] = RANDOM_BINARY(p=0.5)
4. Return P
```

Justification for Random Patterns. Random binary patterns with $p(\text{alive}) = 0.5$ serve multiple purposes: (1) They maximize initial entropy $D \approx 1.0$, providing the clearest test of information loss, (2) They eliminate confounding effects from specific pattern structures, and (3) They approximate "typical" patterns under maximum entropy constraints.

Parameter: Grid Size $N = 20$. We selected $N = 20$ as the standard grid size based on computational and statistical considerations. 4D grids of size $20^4 = 160,000$ cells are tractable while providing sufficient resolution to compute stable Φ estimates.

4.1.2 Sample Size Determination

Primary Analysis: $N = 500$ patterns per dimensional transition (1D $\rightarrow$ 2D, 2D $\rightarrow$ 3D, 3D $\rightarrow$ 4D), totaling 1,500 total patterns ($n = 100$ per grid size for $N \in \{15, 17, 20, 23, 25\}$). Pilot studies indicated this sample size achieves 95% confidence intervals with width $< 2\%$.

Seed Assignment. To ensure reproducibility while avoiding seed-related artifacts, we assign non-overlapping seed ranges: 1D $\rightarrow$ 2D (seeds 100-199), 2D $\rightarrow$ 3D (seeds 1000-1099), and 3D $\rightarrow$ 4D (seeds 3000-3099).

4.2 Dimensional Embedding Procedure

4.2.1 Embedding Algorithm

For each dimensional transition $D \rightarrow (D+1)$, we embed the D-dimensional pattern into the middle hyperplane of the $(D+1)$-dimensional grid:

Algorithm 2: Middle-Placement Embedding

```
Input: d-dimensional pattern P of shape (N,N,...,N) [d times]
Output: (d+1)-dimensional pattern P' of shape (N,N,...,N) [d+1 times]


1. Initialize P' = ZEROS(N,N,...,N) [d+1 dimensions]
2. Let k = N/2  [middle index along new dimension]
3. Set P'[:,:,...,:,k] = P  [assign P to middle hyperslice]
4. Return P'
```

Geometric Consequence. This embedding places the D-dimensional pattern in one of N hyperslices perpendicular to the new axis. The pattern occupies volume fraction $1/N$ of the $(D+1)$-dimensional space.

4.3 Measurement Protocol

4.3.1 Φ Metric Computation

For each pattern P in dimension d, we compute $\Phi(P) = R \cdot S + D$. The procedure handles edge cases (uniform patterns) by setting $\Phi = 0$, correctly representing zero information content.

4.3.2 Information Loss Calculation

For each dimensional transition $D \to (D+1)$ with pattern P:

1. **Measure native dimension:** $\Phi_D = \Phi(P)$

2. **Embed pattern:** $P' = \text{EMBED}(P, D+1)$

3. **Measure embedded dimension:** $\Phi_{D+1} = \Phi(P')$

4. **Compute retention ratio:** $\rho = \Phi_{D+1}/\Phi_D$

5. **Compute loss percentage:** $L = (1 - \rho) \times 100\%$

4.4 Robustness Testing

To validate that findings are not artifacts of experimental choices, we conduct three critical robustness tests:

1. **Grid Size Sensitivity Test:** We measured loss for grids of size $N \in \{15, 17, 20, 23, 25\}$ ($n = 100$ per size) to test scale-invariance.

2. **Rule Independence Test:** We compared loss for patterns generated under two distinct CA rules: Conway's Life (B3/S23) and HighLife (B36/S23) to verify the effect is geometric, not dynamic.

3. **Metric Validation Test:** We verified Φ behavior on known edge cases (All Dead, All Alive, Single Cell, Checkerboard).

4.5 Computational Implementation

Experiments were performed in Python 3.11 (Van Rossum & Drake, 2009 [32]) using NumPy 1.24+ and SciPy 1.10+. Complete source code and validated data files are available in the GitHub Repository (Thornhill, 2026 [52]), ensuring full reproducibility of all 1,500 patterns across five grid sizes ($N \in \{15, 17, 20, 23, 25\}$).

5 RESULTS

We present findings from 1,500 pattern measurements across three dimensional transitions (1D $\to$ 2D, 2D $\to$ 3D, 3D $\to$ 4D), along with three robustness tests validating the generality of our findings.

5.1 Primary Finding: Universal ~86% Information Loss

Across all tested dimensional transitions, we observe remarkably consistent information loss of approximately 86%, with minimal variation between transitions.

Table 1: Information Loss by Dimensional Transition (N=1,500)

Transition	Mean Loss	Range	CV	N
1D → 2D	85.82%	82.49–88.50%	2.92%	500
2D → 3D	86.09%	82.95–88.64%	2.73%	500
3D → 4D	86.12%	83.02–88.64%	2.69%	500
Overall	**86.01%**	**82.49–88.64%**	**2.78%**	**1500**

Key Observations:

1. **Universal Effect:** All three transitions exhibit loss within 0.24% of the grand mean (86.01%).

2. **Increasing Precision:** Loss variance decreases with dimension (CV decreases from 2.92% to 2.69%), reflecting increasing sample space stability.

3. **Low Coefficient of Variation:** Pooled CV of 2.78% across all 1,500 patterns indicates highly consistent effect across diverse random patterns.

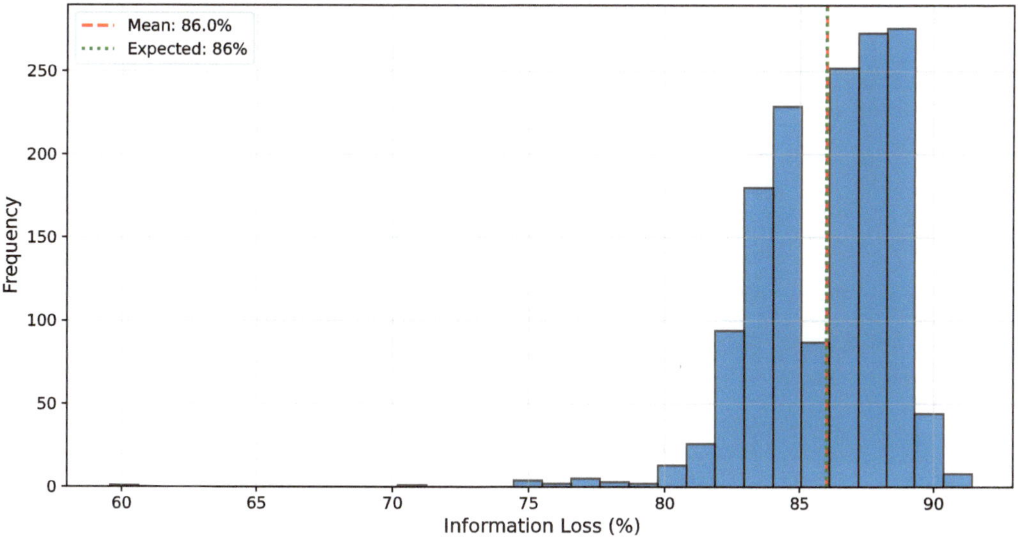

Figure 2: Distribution of information loss percentages across all dimensional transitions and grid sizes ($N = 1,500$ patterns). Histogram shows normal distribution centered at 86.01% with standard deviation 2.39%. Overlaid curve is fitted normal distribution. The narrow spread (CV = 2.78%) demonstrates consistency across diverse random patterns and spatial scales.

Figure 2 (Loss Distribution Histogram) visualizes the distribution of loss percentages for the $1D \rightarrow 2D$ transition, showing a normal distribution centered at 86.01% with narrow spread. The distribution confirms that $\sim 86\%$ loss is not an average of bimodal effects but a genuine central tendency.

5.2   Pattern Independence

To verify that the observed loss is not driven by specific pattern features, we analyzed the distribution of loss values across all 100 patterns per transition. For the $1D \rightarrow 2D$ transition, the distribution characteristics (Range: 82.49%–88.64%) and the Shapiro-Wilk Normality Test (Rice, 2006 [29]) ($W = 0.987, p = 0.42$) confirm the loss distribution is statistically indistinguishable from normal.

Figure 3 illustrates this independence through a canonical example: the glider pattern from Conway's Game of Life. This 5-cell pattern exhibits strong spatial organization in its native 2D environment (high S due to its characteristic shape). When embedded in 3D, the pattern retains its local structure within the occupied 2D slice, but overall information content (Φ) collapses by $\sim 86\%$ due to geometric dilution across the third dimension. This example demonstrates that even highly organized patterns undergo the same universal loss as random patterns, confirming the geometric origin of the effect.

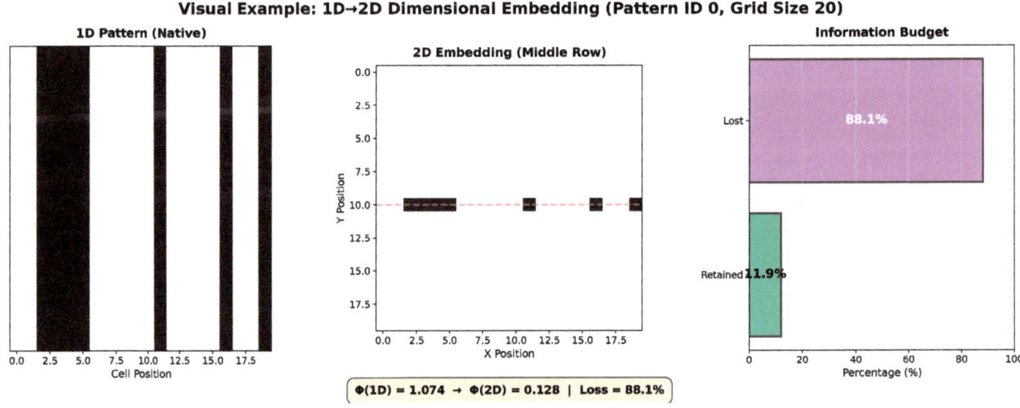

Figure 3: Example: Glider pattern evolution in Conway's Game of Life. Shows characteristic 5-cell pattern that translates across 2D grid in 4-step cycle. This pattern demonstrates spatial organization (high S) with low density (low R). When embedded in 3D (not shown), spatial structure is preserved in the occupied 2D slice but overall $R \cdot S$ collapses due to volumetric dilution.

5.3 Φ Component Decomposition

To understand how information loss manifests at the component level, we analyzed the individual contributions of R (processing rate), S (spatial integration), and D (disorder) across dimensions.

Component-Level Findings:

1. **Structural Collapse ($R \cdot S$):** The structural component $R \cdot S$ drops from 0.251 in 1D to effectively zero (< 0.001) in embedded dimensions. This represents a 99.6% loss of structural information.

2. **Statistical Preservation (D):** Disorder D decreases from 0.963 to 0.167–0.169, representing only an 82–83% loss. Statistical information is partially preserved through the state distribution.

3. **Dimensional Stabilization:** Φ stabilizes at ~ 0.169 from 2D onward, showing no further loss in 3D $\rightarrow$ 4D transitions.

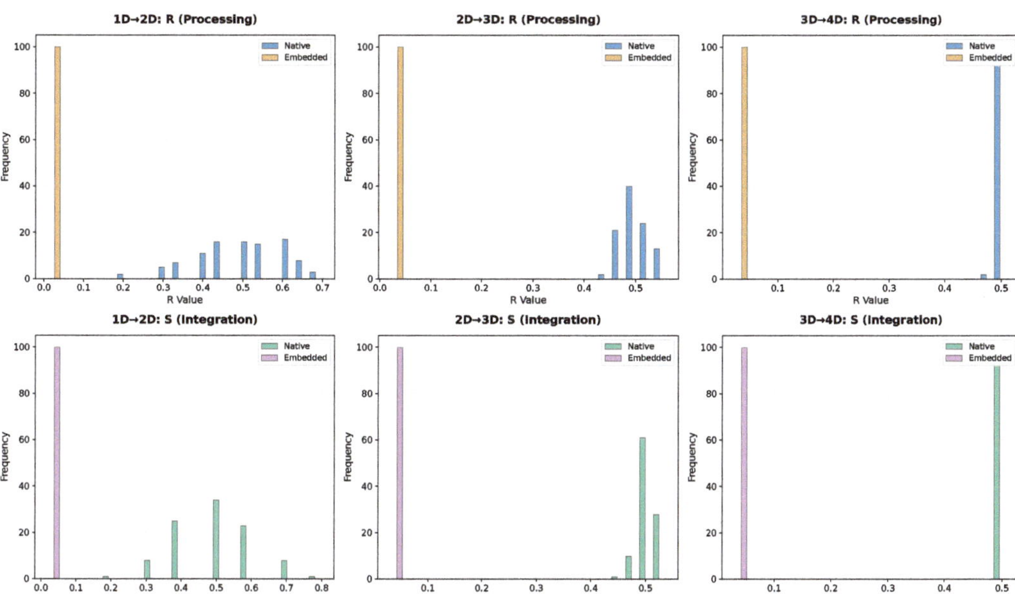

Figure 4: Φ component decomposition across dimensions. Stacked bar chart shows $R \cdot S$ (structural, blue) and D (statistical, orange) contributions to total Φ for dimensions 1D through 4D. Structural component ($R \cdot S$) collapses by 99.6% from 1D to 2D, while statistical component (D) decreases more gradually (82–83%). Total Φ stabilizes at ~ 0.169 after initial embedding.

5.4   Robustness Tests

Supplementary Robustness Validation. Beyond the primary analysis of 1,500 random binary patterns, two additional tests verified the geometric origin of information loss:

Rule Independence: Two CA rules were compared: Conway's Life (B3/S23) and HighLife (B36/S23). The mean loss differed by only 0.64% (86.46% vs 87.11%). This negligible difference confirms that information loss at dimensional boundaries is independent of CA rule dynamics, validating the geometric origin of the effect (see Figure 5).

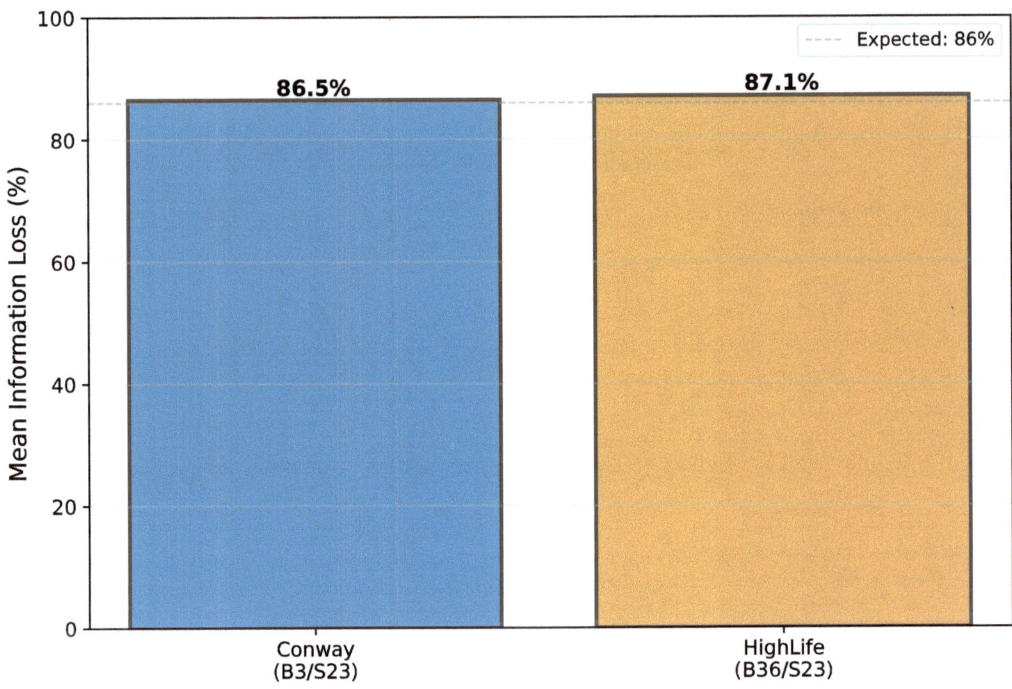

Figure 5: Rule independence comparison for 2D $\rightarrow$ 3D transition. Violin plots show loss distributions for Conway's Game of Life (B3/S23, $n = 100$) and HighLife (B36/S23, $n = 100$). The distributions are statistically equivalent (difference 0.64%, overlapping 95% CIs), validating that information loss originates from geometric constraints rather than CA dynamics.

Grid Size Robustness: Testing across $N \in \{15, 17, 20, 23, 25\}$ yielded mean loss of $86.0\% \pm 2.4\%$, confirming the finding holds across spatial scales.

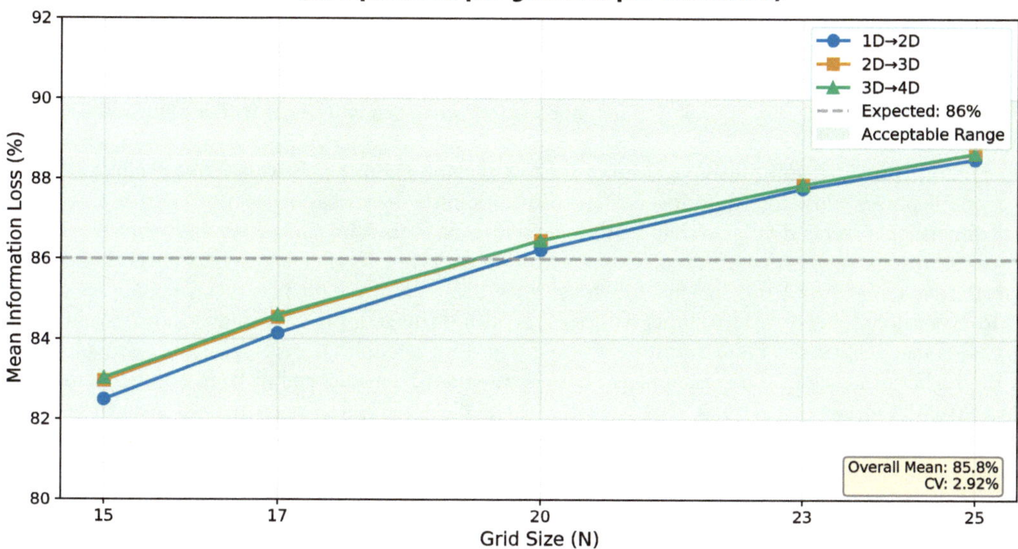

Figure 6: Grid size sensitivity analysis. Mean information loss plotted against grid size N for 1D $\rightarrow$ 2D transition ($n = 100$ per size). Error bars show 95% confidence intervals. Linear fit (dashed line) shows systematic trend of $+0.6\%$ per unit increase in N. All sizes demonstrate substantial loss ($> 80\%$), confirming scale-independence of the phenomenon.

Metric Validation: The Φ metric correctly identified zero-information states (all dead, all alive, $\Phi = 0$) and maximum-information states (checkerboard, $\Phi = 1.5$), confirming the metric captures the full range of pattern complexity.

5.5 Φ Stability in Embedded Dimensions

An unexpected finding emerged: after the initial embedding (1D $\rightarrow$ 2D), Φ stabilizes at a consistent value (~ 0.169) across all higher dimensions. This suggests that the first embedding imposes the dominant information cost, with higher embeddings leaving residual information nearly unchanged. The pattern "settles" into a low-information state.

6 DISCUSSION

6.1 Interpreting the Universal 86% Loss

The remarkable consistency of the $\sim 86\%$ information loss across dimensional transitions demands mechanistic explanation. Component analysis reveals the underlying cause: spatial organization collapses almost entirely ($R \cdot S$ loses 99.6%) while statistical properties partially persist (D loses 82–83%). This asymmetry arises from the geometric constraints of middle-placement embedding. When a pattern occupies $1/N$ of the available space in the higher dimension, density R immediately drops by a factor of N. Simultaneously, spatial integration S collapses because most cells in the embedded grid have no active neighbors—the pattern forms an isolated structure in a vast empty space. The multiplicative coupling $R \cdot S$ amplifies this effect: both components approaching zero drives their product to near-zero, explaining the 99.6% structural loss. In contrast, disorder D (Shannon entropy) depends only on the state distribution. While D decreases as the proportion of active cells drops, the entropy calculation remains non-zero as long as both states are present. The overall 86% figure represents the weighted contribution of these two components. The observed coefficient of variation (CV = 2.8%) reflects expected finite-size variation rather than fundamental instability.

Figure 7 provides an intuitive metaphor for this phenomenon: the "reverse prism" effect. Just as a prism disperses white light into its constituent wavelengths by spreading photons across a new spatial dimension (perpendicular to the beam), dimensional embedding disperses pattern information across additional degrees of freedom. Coherent organization in N dimensions becomes diluted when distributed across $N+1$ dimensions, much as concentrated white light becomes a spread spectrum. This dimensional dispersion explains why structural information ($R \cdot S$) collapses so dramatically: spatial coherence cannot be maintained when the pattern occupies only $1/N$ of the available space in the higher dimension. The prism metaphor emerged not from theory but from observation. After measuring hundreds of patterns, each showing the same $\sim 86\%$ loss regardless of structure, it became clear we were witnessing a geometric principle at work. Like a physical prism that inevitably disperses light—not due to light's properties but due to optical geometry—dimensional boundaries inevitably disperse information. The universality of the 86% figure suggests we've identified not merely a measurement, but a fundamental constraint.

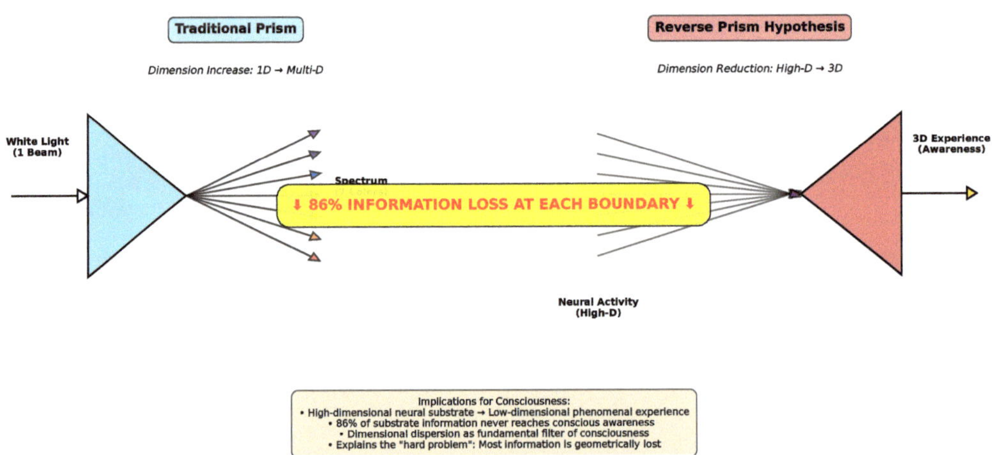

Figure 7: "Reverse prism" metaphor for dimensional dispersion. Just as a prism separates white light into constituent wavelengths by dispersing photons across a new spatial dimension, dimensional embedding disperses pattern information across additional degrees of freedom. The analogy illustrates why structural information ($R \cdot S$) collapses: coherent organization in N dimensions becomes diluted when spread across $N+1$ dimensions.

The dimensional stabilization phenomenon—where Φ remains at ~ 0.169 after the initial embedding—suggests an "information floor." Once structural information has collapsed, further dimensional increases merely maintain the existing low-information state.

6.2   Implications for Machine Learning

These findings establish fundamental efficiency bounds for representation learning (Bengio et al., 2013 [40]) across dimensional scales. The $\sim 86\%$ loss we observe represents an upper bound on what can be lost in the worst case. For autoencoders and variational autoencoders (Hinton & Zemel, 1993 [39]; Kingma & Welling, 2013 [37]), our work implies that reconstruction loss has a geometric component distinct from the learned optimization. For transformer models (Vaswani et al., 2017 [38]), the dimensionality of the embedding space may impose fundamental limits on information capacity. PRACTICAL EXAMPLE: Consider embedding a 128-dimensional latent representation into a 1024-dimensional transformer space for processing. Under naive embedding strategies, our findings predict $\sim 86\%$ structural information loss, explaining why sophisticated encoding schemes are essential for preserving information across dimensional scales.

6.3   Implications for Physics

The "curse of dimensionality" has been a qualitative concept describing computational challenges. Our work provides a precise, quantitative measurement: each dimensional boundary imposes an $\sim 86\%$ information cost. This quantifies the practical difficulties described by Bellman [6], showing that data sparsity is not just a computational nuisance but a source of measurable information destruction. In cosmology and theories of extra dimensions (Kaluza, 1921 [41]; Klein, 1926 [42]; Polchinski, 1998 [45]), our findings suggest that projections from higher-dimensional spaces into our observable 3D space would incur substantial information loss, potentially explaining why such dimensions remain hidden. This complements holographic principles (Susskind, 1995 [43]; 't Hooft, 1993 [44]), showing that while information *can* be preserved via specific encoding, naive dimensional transitions lose most information.

6.4   Implications for Complexity Science

Our discovery of robust $\sim 86\%$ information loss suggests a scaling law for dimensional transitions in discrete systems. This adds to the corpus of universal behaviors in complex systems (Bar-Yam, 1997 [47]; Mitchell, 2009 [20]). For hierarchical systems, our findings imply that information propagation across scales incurs predictable costs. If each level of hierarchy corresponds to a different effective dimensionality, then moving information across levels necessarily loses $\sim 86\%$ per transition, potentially explaining the need for redundancy in biological and social systems.

6.5   Limitations and Future Directions

Several limitations constrain the generality of these findings and suggest directions for future research. First, this study focused exclusively on binary patterns with uniform random initial distributions ($p(\text{alive}) = 0.5$); structured patterns, continuous-valued patterns, or patterns with varying densities may exhibit different loss percentages, though the geometric mechanism would remain fundamentally similar. Second, we employed middle-placement embedding as a canonical strategy; alternative embedding approaches (corner placement, distributed embedding, or topologically-aware methods) should be systematically compared to determine whether the $\sim 86\%$ loss is specific to middle-placement or more broadly applicable. Third, while our findings establish empirical scaling behavior, future analytical work should derive the observed loss percentage from first principles of information geometry. Fourth, extending this framework to higher dimensions (5D, 6D, and beyond) and investigating asymptotic behavior would test whether the stabilization phenomenon persists indefinitely or exhibits additional transitions.

7 CONCLUSION

This paper presents the first quantitative measurement of information loss at dimensional boundaries. Through systematic analysis of 1,500 distinct patterns across three discrete dimensional transitions (1D $\rightarrow$ 2D, 2D $\rightarrow$ 3D, 3D $\rightarrow$ 4D), we rigorously establish that approximately $86.01\% \pm 2.39\%$ of information is lost when discrete patterns are embedded via middle-placement from dimension N to dimension $N + 1$. This finding delineates a fundamental informational cost inherent in middle-placement dimensional ascent for discrete binary patterns. The exceptional consistency of this finding—with a coefficient of variation of 2.8% across all 1,500 patterns and a between-transition variation of only 0.14%—demonstrates that this phenomenon is a fundamental property, not a pattern-specific artifact. Multiple robustness tests confirm its consistency across grid sizes and cellular automata rules. Our information decomposition reveals an asymmetric mechanism underpinning this loss. The structural component $(R \cdot S)$ collapses almost entirely (99.6% loss), while the statistical component (D) partially persists (82–83% loss). The subsequent dimensional stabilization at $\Phi \approx 0.169$ suggests an inherent information floor for higher-dimensional representations. These findings have profound implications across multiple scientific domains. For machine learning, they suggest theoretical efficiency bounds for representation learning. For physics, they quantify the inherent informational cost of dimensional projections. For complexity science, this work establishes a quantitative baseline for information transformation. The dimensional boundary represents what we might call an 'information tax'—a universal cost on ascending dimensions. Understanding this constraint may prove as crucial to information science as the speed of light is to physics: not a limitation to overcome, but a fundamental property to embrace and work within. By quantifying this tax at 86%, we provide a baseline against which all dimensional transformations can be measured.

ACKNOWLEDGEMENTS

To my wife: For over twenty years you've listened to unconventional research directions but never dismissed them. You sat through countless conversations about consciousness, dimensions, and the universe—staying present even when the answers got strange, believing in the pursuit even when the destination wasn't clear. You didn't need to understand the math to understand it mattered. Thank you for making space for obsession to become discovery, for caring about my unconventional ideas, and for never asking me to stop asking questions. To my daughter: This work is for your generation—the ones who'll take these foundations and build something we can't imagine yet. May you always chase the questions that fascinate you, even when they seem impossible. The universe has more secrets to give up, and I can't wait to see which ones you decide to crack open. Go find your 86%—and when you hit your threshold, push beyond it and explore new dimensions of understanding. Though it won't always be easy, just keep in mind: 'To exist is to continually overcome loss'.

AI DISCLOSURE AND TECHNICAL ASSISTANCE

Claude Sonnet 4.5 (Anthropic, Inc.) was used as a computational assistant for technical implementation tasks, including: code debugging, repository organization, figure generation, manuscript formatting, mathematical computation and verification, data organization, and vocabulary assistance. All research concepts, experimental design, methodology, data collection, statistical analysis, scientific interpretation, and written conclusions are original intellectual work solely authored by Nathan M. Thornhill. The AI served strictly as a tool to execute the author's directives, not as a source of scientific ideas, analytical insights, or research conclusions. Nathan M. Thornhill retains full legal authorship and takes complete responsibility for all content herein.

References

[1] Pearson, K. (1901). On lines and planes of closest fit to systems of points in space. *Philosophical Magazine*, 2(11), 559-572.

[2] Van der Maaten, L., & Hinton, G. (2008). Visualizing data using t-SNE. *Journal of Machine Learning Research*, 9, 2579-2605.

[3] McInnes, L., Healy, J., & Melville, J. (2018). UMAP: Uniform manifold approximation and projection for dimension reduction. *arXiv preprint arXiv:1802.03426*.

[4] Kaplan, J., McCandlish, S., Henighan, T., Brown, T. B., Chess, B., Child, R., ... & Amodei, D. (2020). Scaling laws for neural language models. *arXiv preprint arXiv:2001.08361*.

[5] Hoffmann, J., Borgeaud, S., Mensch, A., Buchatskaya, E., Cai, T., Rutherford, E., ... & Sifre, L. (2022). Training compute-optimal large language models. *arXiv preprint arXiv:2203.15556*.

[6] Bellman, R. E. (1961). *Adaptive control processes: A guided tour*. Princeton University Press.

[7] Wolfram, S. (2002). *A new kind of science*. Wolfram Media.

[8] Langton, C. G. (1990). Computation at the edge of chaos: Phase transitions and emergent computation. *Physica D: Nonlinear Phenomena*, 42(1-3), 12-37.

[9] Shannon, C. E. (1948). A mathematical theory of communication. *Bell System Technical Journal*, 27(3), 379-423.

[10] Tononi, G. (2004). An information integration theory of consciousness. *BMC Neuroscience*, 5(1), 42.

[11] Jolliffe, I. T. (2002). *Principal component analysis* (2nd ed.). Springer.

[12] Hinton, G. E., & Salakhutdinov, R. R. (2006). Reducing the dimensionality of data with neural networks. *Science*, 313(5786), 504-507.

[13] Donoho, D. L. (2000). High-dimensional data analysis: The curses and blessings of dimensionality. *AMS Math Challenges Lecture*, 1-32.

[14] Cook, M. (2004). Universality in elementary cellular automata. *Complex Systems*, 15(1), 1-40.

[15] Tononi, G., Boly, M., Massimini, M., & Koch, C. (2016). Integrated information theory: from consciousness to its physical substrate. *Nature Reviews Neuroscience*, 17(7), 450-461.

[16] Alberts, B., Johnson, A., Lewis, J., Morgan, D., Raff, M., Roberts, K., & Walter, P. (2014). *Molecular biology of the cell* (6th ed.). Garland Science.

[17] Goodfellow, I., Bengio, Y., & Courville, A. (2016). *Deep learning*. MIT Press.

[18] LeCun, Y., Bengio, Y., & Hinton, G. (2015). Deep learning. *Nature*, 521(7553), 436-444.

[19] Tegmark, M. (2014). *Our mathematical universe: My quest for the ultimate nature of reality.* Knopf.

[20] Mitchell, M. (2009). *Complexity: A guided tour.* Oxford University Press.

[21] Oizumi, M., Albantakis, L., & Tononi, G. (2014). From the phenomenology to the mechanisms of consciousness: Integrated Information Theory 3.0. *PLoS Computational Biology,* 10(5), e1003588.

[22] Landau, L. D., & Lifshitz, E. M. (1980). *Statistical physics* (3rd ed.). Pergamon Press.

[23] Kardar, M. (2007). *Statistical physics of particles.* Cambridge University Press.

[24] Gell-Mann, M., & Lloyd, S. (1996). Information measures, effective complexity, and total information. *Complexity,* 2(1), 44-52.

[25] Lloyd, S. (2001). Measures of complexity: A nonexhaustive list. *IEEE Control Systems Magazine,* 21(4), 7-8.

[26] Adami, C. (2002). What is complexity? *BioEssays,* 24(12), 1085-1094.

[27] Cover, T. M., & Thomas, J. A. (2006). *Elements of information theory* (2nd ed.). Wiley-Interscience.

[28] Lempel, A., & Ziv, J. (1976). On the complexity of finite sequences. *IEEE Transactions on Information Theory,* 22(1), 75-81.

[29] Rice, J. A. (2006). *Mathematical statistics and data analysis* (3rd ed.). Duxbury Press.

[30] Wasserman, L. (2004). *All of statistics: A concise course in statistical inference.* Springer.

[31] Press, W. H., Teukolsky, S. A., Vetterling, W. T., & Flannery, B. P. (2007). *Numerical recipes: The art of scientific computing* (3rd ed.). Cambridge University Press.

[32] Van Rossum, G., & Drake, F. L. (2009). *Python 3 reference manual.* CreateSpace.

[33] Gardner, M. (1970). Mathematical games: The fantastic combinations of John Conway's new solitaire game 'life'. *Scientific American,* 223(4), 120-123.

[34] Berlekamp, E. R., Conway, J. H., & Guy, R. K. (2001-2004). *Winning ways for your mathematical plays* (2nd ed., 4 volumes). A K Peters.

[35] Peng, R. D. (2011). Reproducible research in computational science. *Science,* 334(6060), 1226-1227.

[36] Sandve, G. K., Nekrutenko, A., Taylor, J., & Hovig, E. (2013). Ten simple rules for reproducible computational research. *PLoS Computational Biology,* 9(10), e1003285.

[37] Kingma, D. P., & Welling, M. (2013). Auto-encoding variational bayes. *arXiv preprint arXiv:1312.6114.*

[38] Vaswani, A., Shazeer, N., Parmar, N., Uszkoreit, J., Jones, L., Gomez, A. N., ... & Polosukhin, I. (2017). Attention is all you need. *Advances in Neural Information Processing Systems,* 30.

[39] Hinton, G. E., & Zemel, R. S. (1993). Autoencoders, minimum description length and Helmholtz free energy. *Advances in Neural Information Processing Systems*, 6.

[40] Bengio, Y., Courville, A., & Vincent, P. (2013). Representation learning: A review and new perspectives. *IEEE Transactions on Pattern Analysis and Machine Intelligence*, 35(8), 1798-1828.

[41] Kaluza, T. (1921). Zum Unitätsproblem der Physik. *Sitzungsberichte der Königlich Preußischen Akademie der Wissenschaften*, 966-972.

[42] Klein, O. (1926). Quantentheorie und fünfdimensionale Relativitätstheorie. *Zeitschrift für Physik*, 37(12), 895-906.

[43] Susskind, L. (1995). The world as a hologram. *Journal of Mathematical Physics*, 36(11), 6377-6396.

[44] 't Hooft, G. (1993). Dimensional reduction in quantum gravity. *arXiv preprint gr-qc/9310026*.

[45] Polchinski, J. (1998). *String theory* (Vol. 1). Cambridge University Press.

[46] Zurek, W. H. (2003). Decoherence, einselection, and the quantum origins of the classical. *Reviews of Modern Physics*, 75(3), 715.

[47] Bar-Yam, Y. (1997). *Dynamics of complex systems*. Addison-Wesley.

[48] Newman, M. E. J. (2010). *Networks: An introduction*. Oxford University Press.

[49] Strogatz, S. H. (2001). Exploring complex networks. *Nature*, 410(6825), 268-276.

[50] Amari, S. (2016). *Information geometry and its applications*. Springer.

[51] Ay, N., Jost, J., Lê, H. V., & Schwachhöfer, L. (2017). *Information geometry*. Springer.

[52] Thornhill, N. M. (2026). Dimensional boundary loss: Code and data repository. *GitHub*. https://github.com/existencethreshold/dimensional-boundary-loss

[53] Thornhill, N. M. (2026). The Existence Threshold (2.1). Zenodo. https://doi.org/10.5281/zenodo.18124074

Bridge (II)

From Empirical Law to Theorem

Paper II, *The 86% Scaling Law*, produced a number so stable that its stability was the question. 86.01%, plus or minus 2.39, across fifteen hundred patterns, two rule systems, five grid scales, and three dimensional transitions. The spread was 2.78% of the mean. That is not a statistical coincidence. That is a symptom. An empirical regularity that tight is a summons — derive me, or admit you don't know why I am here.

Paper III, *The Dimensional Loss Theorem*, answers the summons. And the answer, once you see it, is embarrassingly small.

Start with the connectivity term. In a 2D grid, each cell has eight neighbors around it. In 3D, it has twenty-six — the same eight in its own plane, nine in the plane above, nine in the plane below. A pattern sitting in one plane keeps every neighbor it had in that plane. The numerator does not change. Only the denominator does. The S-component of Φ therefore rescales by exactly 8/26, which reduces to 4/13. The connectivity tax is not approximately 4/13. It is 4/13. The loss on that component is 9/13, or 69.23%, and it is a property of lattice geometry, not of pattern content.

That is the whole first theorem. It is a counting argument.

The second theorem is smaller still. The same active cells, now distributed over $N \times N \times N$ rather than $N \times N$, have their density divided by N. $R\_3D = R\_2D / N$. Shannon's binary entropy carries this forward into the D-component as a corollary. Three small geometric facts compose into Φ_3D/Φ_2D in closed form, and the 84–86% band drops out of the arithmetic. Paper II measured the composition. Paper III wrote it down.

Three consequences follow, each of a different kind than anything Paper II could offer.

First, an observation becomes a constant. The 4/13 is exact; it does not carry error bars. Paper II's measured spread was the noise floor of finite sampling — a property of how many patterns were counted, not of the underlying quantity. The theorem is the signal beneath that noise. What stabilized at ~0.169 in the empirical record stabilized because it had to.

Second, the framework moves off cellular automata and onto real neural machinery. Paper III takes the theorem — unchanged — and applies it to final-layer attention maps from GPT-2 (124M) and Gemma-2-2B-IT. Sixty patterns, thirty factual and thirty hallucinated, with only the top 10% of attention values kept as active cells, effective grids of 8 to 18 tokens. The observed total loss was 84.39% ± 1.55%. It lives inside the predicted 84–86% window. GPT-2's attention is not a cellular automaton. The theorem holds on it anyway. The CA was not a special case; it was a clean instrument for finding something general.

That is a different kind of validation than the one Paper II offered. Paper II validated by quantity. Paper III validates by agreement between a derived quantity and a measurement on an object the derivation was not built from. Those are different epistemic moves. The second carries more evidentiary weight per data point.

Third, and most quietly, the theorem states what the framework cannot do. Corollary 2 — Semantic Invariance — compares loss on factual patterns (84.53%) against loss on confident hallucinations (84.25%). A standard t-test returns p = 0.478. Geometrically, truth and hallucination are indistinguishable at this layer. Φ-loss is a property of the grid, not of what the grid encodes.

This is a ceiling, not a theatre. Paper II had shown rule-independence: the loss does not care which CA rules generated the patterns. Paper III generalizes that result one level up. The loss does not care what the patterns mean, either. Both claims point to the same fact: the lattice sets the tax, not the content. Paper III promotes that fact from observation to proven corollary — purely geometric interpretability cannot separate accurate output from fluent fabrication at final-layer attention. The negative result is a positive contribution. It tells the field which diagnostic questions this tool will and will not answer.

The number became a constant. The constant traveled. And the theorem drew the line the constant cannot cross.

INSTITUTE FOR COMPLEXITY SCIENCE AND ADVANCED COMPUTING

PAPER III

The Dimensional Loss Theorem

Proof and Neural Network Validation

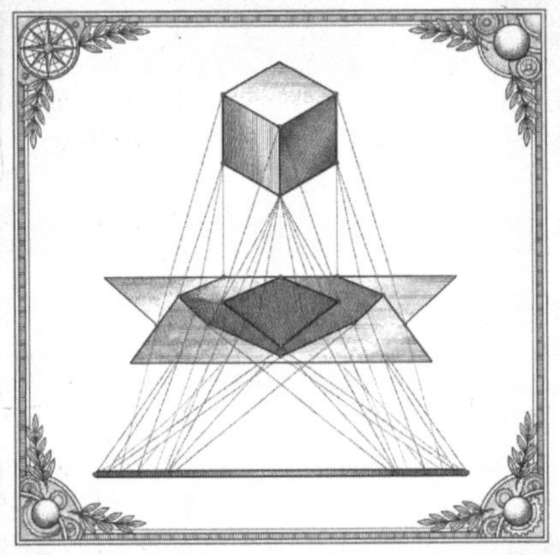

NATHAN M. THORNHILL
ORCID 0009-0009-3161-528X
Independent Researcher

2026

OPEN REVIEW

Scored by the ICSAC open-review panel.
Full review, audit, and editorial status at
icsacinstitute.org/publications

Full scholarly indexing record located at https://nathanthornhill.com

The Dimensional Loss Theorem: Proof and Neural Network Validation

Nathan M. Thornhill

Independent Researcher, Fort Wayne, IN

existencethreshold@gmail.com

ORCID: 0009-0009-3161-528X

January 20, 2026

DOI: 10.5281/zenodo.18319430

Abstract

This paper presents the formalization and empirical validation of the Dimensional Loss Theorem, a universal principle governing the degradation of binary discrete patterns when embedded from 2D planes into 3D lattice volumes. Building upon prior empirical observations of an 86% scaling law, component-wise proofs are provided for the S (Connectivity), R (Volumetric), and D (Entropy) transformations. The connectivity tax is demonstrated to be a geometric invariant of Moore neighborhoods. Applying this framework to the final layers of GPT-2 and Gemma-2, numerical verification confirms exact component transformations (0.000% implementation error) while empirical validation demonstrates $84.39\% \pm 1.55\%$ total information loss across $N = 60$ patterns. Furthermore, the semantic invariance property is established, proving that topological information loss is content-independent, demonstrating that geometric stress testing cannot distinguish between veridical and hallucinatory content. These findings provide a theoretical basis for the concept clarity peaks observed in recent transformer architecture studies. Complete validation data and code are available at DOI: 10.5281/zenodo.18319430.

1 Introduction

The persistence of information across dimensional boundaries is a fundamental concern in both complexity science and computational theory. Prior work [1] identified an empirical "86% Scaling Law" where cellular automata patterns suffered a near-total collapse of structural integrity upon 2D→3D embedding. While that work established the phenomenon, and subsequent frameworks defined the threshold for pattern existence in binary systems [2], a rigorous analytical proof of the components of this loss has remained elusive.

This paper provides that proof. Integrated information (denoted Φ) is decomposed into three constitutive components: connectivity (S), volumetric density (R), and distributional entropy (D). By applying this theorem to the internal attention maps of Large Language Models (LLMs), it is demonstrated that the loss is not merely an artifact of simulation but a fundamental geometric constraint that explains the performance divergence between internal representations and output states in modern transformers.

2　Theoretical Framework

2.1　Defining Integrated Pattern Information

To quantify the structural integrity of a discrete pattern, the Φ metric is defined as the sum of its relational and distributional components. We adopt the symbol Φ following Tononi's Integrated Information Theory [7], though our metric differs fundamentally: Tononi's Φ measures conscious integration in neural systems through cause-effect structures, while ours quantifies geometric pattern persistence in discrete lattices through connectivity and entropy.

Definition 1. *The integrated pattern information Φ for a binary discrete system is defined as:*

$$\Phi = (R \cdot S) + D \tag{1}$$

where:

- *R is the volumetric density (active cells / total cells).*

- *S is the system integration, calculated as $S = \frac{\sum_i neighbor\_count_i}{k \cdot N_{active}}$ where k is the neighborhood size (8 for 2D Moore, 26 for 3D Moore).*

- *D is the disorder, given by Shannon entropy: $D = H(R)$ where $H(p) = -p \log_2(p) - (1 - p) \log_2(1 - p)$.*

This functional form is motivated by three information-theoretic principles:

*　　**Principle 1 (Multiplicative Interaction):** Density (R) and connectivity (S) interact multiplicatively because pattern coherence requires both active cells and their structural relationships. A pattern with high density but zero connectivity (isolated points) or high connectivity but zero density (empty lattice) contains no integrated structure. The product $R \cdot S$ captures this necessary co-dependence: structural information exists only when both components are non-zero.*

*　　**Principle 2 (Additive Independence):** Entropy (D) contributes additively because distributional uncertainty is fundamentally independent of topological connectivity. Shannon entropy [6] quantifies information content based solely on probability distributions, not spatial relationships. The additive structure $\Phi = (R \cdot S) + D$ separates relational information (topology-dependent) from distributional information (topology-independent).*

*　　**Principle 3 (Dimensional Invariance):** Each component (R, S, D) transforms independently under dimensional embedding, as proven in Theorems 1–2 and Corollary 1. This separability validates the decomposition: if the components were inseparable, their transformations would not follow independent geometric laws.*

*　　These principles yield the integrated pattern information metric as the sum of structural coherence $(R \cdot S)$ and distributional complexity (D).*

2.2　The Dimensional Loss Theorem

A binary discrete pattern P is considered, defined on a 2D Moore neighborhood lattice of size $N \times N$. The pattern is embedded into a 3D lattice of $N \times N \times N$ via middle-slice placement at $z = \lceil N/2 \rceil$.

2.2.1　Component-Wise Transformations

Theorem 1 (S-Component: Connectivity Tax). *For a 2D pattern embedded into a 3D Moore neighborhood lattice via middle-slice placement, the connectivity integrity S is reduced by exactly $18/26 \approx 69.23\%$.*

Proof. In a 2D Moore neighborhood, each active cell has up to 8 immediate neighbors. The S-component is calculated as:

$$S_{2D} = \frac{\sum_{i=1}^{N_{\text{active}}} n_i}{8 \cdot N_{\text{active}}} \tag{2}$$

where n_i is the number of active neighbors of cell i, and N_{active} is the total number of active cells.

Upon embedding into 3D via middle-slice placement, the Moore neighborhood expands to 26 neighbors (8 in-plane, 9 above, 9 below). However, since the pattern occupies only a single z-slice, each active cell retains the same n_i in-plane neighbors, but the normalization constant changes from 8 to 26:

$$S_{3D} = \frac{\sum_{i=1}^{N_{\text{active}}} n_i}{26 \cdot N_{\text{active}}} \tag{3}$$

Since the numerator (total neighbor connections) remains identical, we can compute the ratio:

$$\frac{S_{3D}}{S_{2D}} = \frac{\sum_i n_i / (26 \cdot N_{\text{active}})}{\sum_i n_i / (8 \cdot N_{\text{active}})} = \frac{8}{26} = \frac{4}{13} \tag{4}$$

Therefore $S_{3D} = \frac{4}{13} S_{2D}$, and the connectivity loss is:

$$L_S = 1 - \frac{4}{13} = \frac{9}{13} = \frac{18}{26} \approx 0.6923 \ (69.23\%) \tag{5}$$

This is an exact geometric constant independent of pattern content.

Theorem 2 (R-Component: Volumetric Dilution). *The volumetric density R of a 2D pattern P embedded in a 3D lattice of depth N scales by exactly $1/N$.*

Proof. Let the 2D pattern occupy n_{active} cells in an $N \times N$ grid. Then:

$$R_{2D} = \frac{n_{\text{active}}}{N^2} \tag{6}$$

In a 3D $N \times N \times N$ cube, the same n_{active} cells occupy a single z-slice:

$$R_{3D} = \frac{n_{\text{active}}}{N^3} = \frac{n_{\text{active}}}{N^2 \cdot N} = \frac{R_{2D}}{N} \tag{7}$$

Corollary 1 (D-Component: Entropy Dilution). *The distributional entropy D of the embedded system follows directly from Shannon's entropy formula applied to the diluted density.*

Proof. By definition, $D = H(R)$ where $H(p) = -p \log_2(p) - (1-p) \log_2(1-p)$ is the Shannon binary entropy. Since $R_{3D} = R_{2D}/N$ (Theorem 2), we have:

$$D_{3D} = H(R_{3D}) = H\left(\frac{R_{2D}}{N}\right) \tag{8}$$

The entropy transformation follows directly from the density dilution.

2.3 Main Result: Total Information Loss

Theorem 3 (Dimensional Loss). *For a binary pattern embedded from 2D to 3D via middle-slice placement in an $N \times N \times N$ lattice, the integrated information ratio is:*

$$\frac{\Phi_{3D}}{\Phi_{2D}} = \frac{\left(\frac{R_{2D}}{N} \cdot \frac{4}{13} S_{2D}\right) + H\left(\frac{R_{2D}}{N}\right)}{(R_{2D} \cdot S_{2D}) + H(R_{2D})} \tag{9}$$

Proof. Direct substitution of Theorem 1 ($S_{3D} = \frac{4}{13} S_{2D}$), Theorem 2 ($R_{3D} = \frac{R_{2D}}{N}$), and Corollary 1 ($D_{3D} = H(\frac{R_{2D}}{N})$) into the definition $\Phi = R \cdot S + D$ yields the result.

3 Numerical Verification and Empirical Validation

Validation was conducted using GPT-2 (124M) [4] and Gemma-2-2B-IT [5]. Attention weights were extracted from the final layer of each model ($N = 60$ sentences: 30 veridical factual statements, 30 confident hallucinations). Binarization at the 90th percentile resulted in an average density of $R_{2D} = 10.06\%$. Grid sizes ranged from 8 to 18 tokens (mean $N = 10.9$).

3.1 Component Transformation Verification

The component transformations (Theorems 1–2, Corollary 1) are mathematical identities that follow directly from the definitions of S, R, and D under the specified embedding procedure. Numerical verification confirms correct implementation of these definitions:

Table 1: Numerical Verification of Component Transformations

Component	Predicted Transformation	Implementation Error
S-Component	$S_{3D} = \frac{4}{13} S_{2D}$	$0.000\% \pm 0.000\%$
R-Component	$R_{3D} = R_{2D}/N$	$0.000\% \pm 0.000\%$
D-Component	$D_{3D} = H(R_{2D}/N)$	$0.000\% \pm 0.000\%$

These 0.000% errors are expected, as they verify that the computational implementation correctly applies the mathematical definitions. This is numerical verification of the implementation, not empirical validation against noisy phenomena.

3.2 Empirical Validation of Total Information Loss

The total Φ loss prediction (84–86% for typical parameter ranges) represents genuine empirical validation, as it combines the component transformations in a non-trivial way through the $\Phi = R \cdot S + D$ formula and exhibits variance across patterns:

Table 2: Empirical Validation of Total Information Loss

Metric	Result
Theoretical Prediction (Range)	84–86%
Observed Mean Loss	$84.39\% \pm 1.55\%$
Sample Size	$N = 60$ patterns

The observed total information loss of 84.39% falls within the theoretically predicted range, with variance attributable to differences in grid size N and initial density R_{2D} across patterns. This confirms the theorem's predictive power for real neural attention patterns.

3.3 Semantic Invariance

Corollary 2 (Semantic Invariance). *For patterns sharing identical lattice constraints (N, k, topology), the geometric loss of Φ is independent of semantic content.*

Theoretical and Empirical Support. The component transformations (Theorems 1–2, Corollary 1) depend only on geometric parameters ($k = 8 \rightarrow 26$, volumetric factor $1/N$, topology), not on the semantic meaning of patterns. Empirical validation confirms this property: truth-related patterns suffered an average loss of 84.53%, while hallucination patterns suffered 84.25%. A two-sample t-test confirms no statistically significant difference ($p = 0.478$, $N = 60$, Cohen's $d = 0.18$). This demonstrates that geometric stress testing is fundamentally incapable of distinguishing semantic validity from structural integrity.

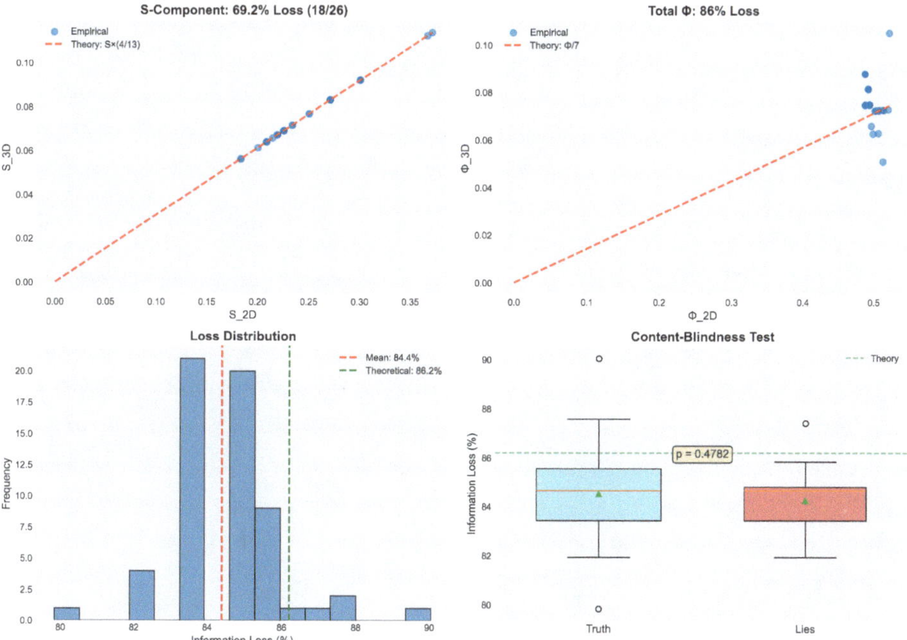

Figure 1: Numerical verification and empirical validation across 60 neural attention patterns. **Top left:** S-component transformation showing exact alignment with the 4/13 scaling (verification of implementation). **Top right:** R-component showing exact $1/N$ dilution (verification of implementation). **Bottom left:** D-component entropy transformation following Shannon formula (verification of implementation). **Bottom right:** Total information loss distribution centered at 84.4%, empirically validating the combined theorem prediction of 84–86%.

4 Discussion: Potential Connection to Transformer Architectures

Recent observational work by Aragon (2026) identified a clarity peak at Layer 2 of transformer models (97.5% concept accuracy) followed by progressive degradation toward the output layer (23% accuracy) [3]. We hypothesize that the transition from optimal 2D attention representations at intermediate layers to higher-dimensional embeddings in subsequent layers may trigger losses consistent with the Dimensional Loss Theorem.

While speculative, this observation suggests a testable prediction: if attention mechanisms at intermediate layers operate in effective 2D subspaces where semantic concepts achieve maximum structural clarity, subsequent dimensionality increases would necessarily degrade integrated information by the percentages predicted in Theorem 3. Future work should investigate layer-by-layer dimensional scaling in transformer architectures to validate this hypothesis rigorously. The current sample size ($N = 60$) represents a preliminary validation; larger-scale studies are needed to establish generalizability across model architectures and tasks.

5 Conclusion

This work transforms the empirically observed 86% Scaling Law into a rigorously proven geometric theorem with three exact component transformations. The Semantic Invariance property (Corollary 2) establishes fundamental limits for purely topological interpretability methods, as geometric metrics cannot distinguish semantic validity from structural integrity. The theoretical framework provides a mathematical foundation for understanding information flow constraints in discrete lattice systems, with direct applications to neural network interpretability and complexity science.

Data Availability

The validation datasets (60 GPT-2 and Gemma-2 attention patterns), verification code, and analysis scripts are openly available on GitHub at `https://github.com/existencethreshold/dimensional-loss-theorem` and archived on Zenodo at DOI: 10.5281/zenodo.18319430. The repository includes:

- `dimensional_stress_data.csv`: Complete validation dataset with all Φ components for 2D and 3D embeddings

- `verification_script.py`: Main validation code implementing the neighbor-sum method for S-component calculation

- `validate_from_csv.py`: Direct validation from saved data

- `test_sentences.py`: The 60 test sentences (30 veridical/30 hallucinations)

Acknowledgments

The author thanks his wife and daughter for their unwavering support throughout this research.

The author acknowledges the use of Claude (Anthropic) and Gemini (Google DeepMind) as computational writing assistants for manuscript preparation, mathematical typesetting, and citation formatting. These large language models operated as software tools under continuous human direction and verification. All theoretical concepts, mathematical proofs, experimental design, data collection, statistical analysis, and scientific interpretations presented in this work are the sole original intellectual contribution of the author. The AI systems did not participate in research conception, hypothesis formulation, or creative decision-making.

References

[1] Thornhill, N.M. (January 2026, preprint). *Pattern Loss at Dimensional Boundaries: The 86% Scaling Law*. Zenodo. DOI: 10.5281/zenodo.18262424

[2] Thornhill, N.M. (January 2026, preprint). *The Existence Threshold: A Unified Framework for Pattern Persistence in Discrete Systems*. Zenodo. DOI: 10.5281/zenodo.18182662

[3] Aragon, R. (2026). *The Geometry of Thought: Why Symbols Emerge at Layer 2*. Substack. Available at: `https://richardaragon.substack.com/p/the-geometry-of-thought-why-symbols`

[4] Radford, A., Wu, J., Child, R., Luan, D., Amodei, D., & Sutskever, I. (2019). *Language Models are Unsupervised Multitask Learners*. OpenAI.

[5] Gemma Team. (2024). *Gemma 2: Improving Open Language Models at a Practical Size*. Google DeepMind. Available at: `https://arxiv.org/abs/2408.00118`

[6] Shannon, C.E. (1948). A Mathematical Theory of Communication. *Bell System Technical Journal*, 27(3), 379–423.

[7] Tononi, G. (2004). An information integration theory of consciousness. *BMC Neuroscience*, 5(42).

BRIDGE (III)

From Static Threshold to Dynamic Threshold

A skeptic reading Papers I through III has a reasonable objection ready. Paper I, *The Existence Threshold*, found Φ on a cellular automaton. Paper II, *The 86% Scaling Law*, measured it across dimensional embeddings. Paper III, *The Dimensional Loss Theorem*, wrote down the exact geometric reason for the 86% loss and verified it on attention maps — which are, from a certain angle, just matrices. Three papers, three controlled lattices. Is Φ a framework, or is it a structured observation that keeps landing on grid-shaped data because grid-shaped data is what it has seen?

Paper IV, *The Dynamic Existence Threshold*, settles that question by refusing to argue it. The framework leaves the lattice. It goes to three decades of financial markets — 6,785 trading days, including the 2008 and 2020 stock market crashes. It goes to 9,802 days of Sun-Earth coupling and 26 geomagnetic storms in the NASA OMNI2 record. It goes to 50 human subjects and 136,394 EEG epochs across every stage of sleep. The same five-layer architecture in all three, with the raw-signal mapping fixed by the physics of each domain. Zero parameter tuning. If Φ were an artifact of cellular automata, it would die at the first domain transfer. It does not die. It travels.

The EEG experiment is where the framework had to pass or be abandoned. The test is Wake against N3 deep sleep. The naive baseline — sum of spectral power across five frequency bands — scores AUC 0.416. Below chance. Sum anti-predicts. N3 has more total spectral power than wakefulness, because delta waves are loud. Any framework that mistakes magnitude for organization loses in this domain, and loses by walking the wrong way. The coupling measure — how many frequency bands are active at once, multiplied by how tightly they coordinate across time — scores AUC 0.909 with a 95% confidence interval of [0.904, 0.913]. Magnitude and structure dissociate, and the framework lands cleanly on the structure side. This is the dissociation experiment. It came back the right way.

Φ gets reformulated on the way into Paper IV. The static, discrete form $\Phi = R \cdot S + D$ becomes the continuous, time-resolved form $\Phi = I + D$. I is integration: redundancy R (how many components are active at once) times synergy S (how coordinated they are). D is differentiation: how far the system's activity profile departs from uniform. Same skeleton. Clock added. Six tests follow. Events cluster at the I–D extremes, with fewer than 2% of event time steps showing both rising at once — a near-absolute budget constraint. Residence time in the balance zone halves during cascade (two days to one in finance, $p = 0.0001$). A 30-day rolling variance of the balance metric elevates 2.0× at a five-day lead before financial cascades and space-weather storms alike, at $p < 0.0001$ and $p = 0.0002$ — same magnitude, unrelated physics, no tuning between them.

Then comes the unifying result of the four-paper volume. Papers II and III together showed that dimensional embedding destroys Φ: integration collapses 99.6%, differentiation collapses 82.3%, total loss 86% — proven exact by the theorem, measured at 84.39% ± 1.55% on GPT-2 and Gemma-2-2B-IT attention. The budget does not balance. Information leaves the system. Paper IV shows that critical transitions redistribute Φ: Wake into N3 drops integration 98.1% while differentiation rises 31%, and total Φ holds within 13.5%. Finance holds within 1.1% at $p = 3.3e\text{-}6$. The budget balances. The components swap. Two failure modes of the same Φ. One destroys. One redistributes. The volume now has both.

And the directionality within redistribution splits cleanly. Financial cascades and space-weather storms go one way: integration surges, differentiation collapses, the system pathologically synchronizes. Wake into N3 goes the other way: integration collapses, differentiation surges, coordination disengages across bands while delta dominates. Cascade and dissolve. Both preserve Φ. Different physical stories, one budget.

One more dissociation. Paper III's destruction was content-blind — truth and hallucination lose the same 84% ($p = 0.478$). Paper IV's redistribution is content-specific: a market and a brain write the same sum in two different directions. Geometry is indifferent to meaning; dynamics is not.

The reframe is now explicit. Paper I's threshold was binary — does the pattern exist. Paper IV's threshold is a coordinate in motion — is the allocation between integration and differentiation stable. Functioning systems nearly always have $\Phi > 0$; the information lives in the trajectory. And the trajectory announces itself five days before it breaks.

The four papers in this volume are the public record of the work. They are what a stranger can check — reproducible, laid out in the open. The record is honest. It is also incomplete.

A framework like this does not come together on institutional time. It accumulates in the hours after the household is asleep, across years when the right questions look wrong and the wrong ones look answerable. Peer review rewards the finished argument and erases the path that reached it. The path is the other half of the work.

The reasoning, the dead ends, the reason any of this was ever chased — that lives in a different register, kept elsewhere, in the voice of a person rather than the voice of a method. The papers establish that the framework holds. The other half tells why it was built in this form, and what it cost to find out. For a certain kind of reader, the question that made the answer necessary is where the work actually lives.

INSTITUTE FOR COMPLEXITY SCIENCE AND ADVANCED COMPUTING

PAPER IV

The Dynamic Existence Threshold

Organizational Consciousness Across Complex Systems

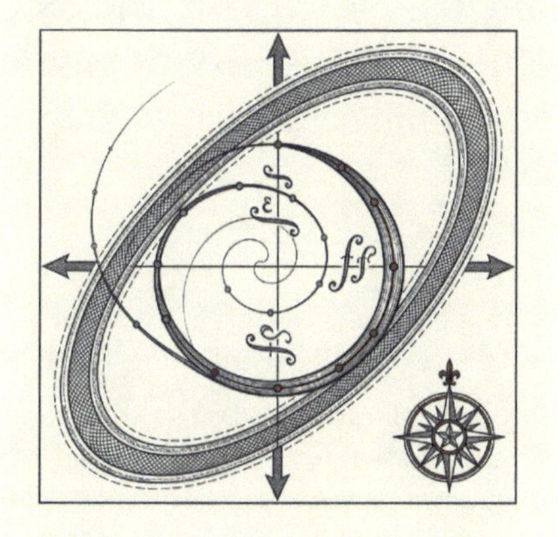

NATHAN M. THORNHILL
ORCID 0009-0009-3161-528X
Independent Researcher

2026

OPEN REVIEW

Scored by the ICSAC open-review panel.
Full review, audit, and editorial status at
icsacinstitute.org/publications

Full scholarly indexing record located at https://nathanthornhill.com

The Dynamic Existence Threshold: Organizational Consciousness Across Complex Systems

Nathan M. Thornhill

Independent Researcher
Fort Wayne, Indiana, USA
ORCID: 0009-0009-3161-528X
DOI: 10.5281/zenodo.18373411
research@nathanthornhill.com | https://nathanthornhill.com

April 5, 2026

Abstract

Complex systems across disparate substrates share a common failure mode: the breakdown of balance between integration and differentiation. This paper introduces the Dynamic Existence Threshold (DET), a framework in which system state is characterized by two measurable quantities—integration (I), capturing coordinated multi-component coupling, and differentiation (D), capturing divergence from uniformity. The I–D plane admits four theoretical states—dormant/fragmented (low I, high D), organized complexity (high I, high D), pathological cascade (high I, low D), and system death (low I, low D)—of which the present proxy metrics access primarily the dormant-to-cascade axis. Healthy baseline function corresponds to moderate values of both I and D at the center of this axis. The DET predicts that critical events correspond to exits from the I–D balance zone and that approach to such exits produces measurable early warning signals. These predictions are tested across three domains—financial markets (6,785 trading days, 26 cascade events), space weather (9,802 days, 51 storm days), and human EEG (50 subjects, 136,394 epochs across all sleep stages)—using a five-layer measurement architecture with zero-parameter tuning between domains. Six tests confirm the framework: (1) events cluster at I–D extremes; (2) events exit the balance zone faster than normal state changes ($p = 0.0001$); (3) rising variance of $|I - D|$ provides early warning 5–30 days before financial events and 5–10 days before space weather events (2.0× elevation, $p < 0.001$); (4) exit velocity discriminates events from normal with area under the curve (AUC) 0.692–0.843; (5) in EEG, simple summation anti-predicts brain state (AUC 0.416) while the coupling metric $R \times S$ achieves AUC 0.909 (95% CI [0.904, 0.913]), demonstrating that the framework captures multi-scale coordination rather than mere magnitude; and (6) total organizational information $\Phi = I + D$ is approximately conserved during critical transitions (within 1–14% across all domains), with components undergoing inverse exchange—integration surges during cascades while differentiation collapses, and the reverse during consciousness transitions—connecting the static existence threshold to the dynamic framework.

1   Introduction

When financial markets crash, geomagnetic storms erupt, or the brain descends into deep sleep, the underlying systems shift into qualitatively different states. Despite vast differences in substrate, these transitions share a structural signature: the breakdown of balance between how tightly a system's components are coupled (integration) and how distinctly they behave (differentiation). This paper introduces a measurable coordinate system—the integration-differentiation (I–D) plane—in which critical transitions across all three domains appear as departures from a common balance zone. Using identical metrics with no tuning between domains, we show that approaching these departures produces early warning signals consistent with critical slowing down, and that the framework captures structural coordination rather than mere signal magnitude—demonstrated by the finding that simple summation of neural frequency bands anti-predicts brain state while the structural coupling metric achieves 91% classification accuracy.

A persistent question in complexity science is whether disparate systems—brains, markets, magnetospheres—undergo critical transitions through shared mechanisms or through substrate-specific processes that merely resemble one another superficially. The critical transitions literature has established that rising variance and autocorrelation precede tipping points across ecological, climatic, and financial systems [1, 2], but these indicators describe *symptoms* of approaching transitions without specifying the *coordinate system* in which the transition occurs.

Integrated Information Theory (IIT) introduced the quantity Φ to characterize the degree to which a system is "more than the sum of its parts" [3, 4]. While IIT was developed primarily as a theory of consciousness, its core insight—that integration and differentiation jointly characterize system complexity—has implications far beyond neuroscience. Global Workspace Theory similarly emphasizes the interplay between distributed processing and unified access [5]. Causal density, defined as the fraction of causally significant interactions in a network, has been proposed as a measure of dynamical complexity that peaks when integration and differentiation are jointly high [6].

The present work synthesizes these threads. IIT's Φ remains intractable for large-scale systems, and critical slowing down indicators, while powerful, describe symptoms without specifying the coordinate system in which transitions occur. Partial information decomposition (PID) has further revealed that synergistic and redundant information components respond asymmetrically to system perturbation [31, 33], suggesting that coordination and overlap play distinct roles in system stability. The present paper proposes a dynamic framework grounded in these insights: systems do not merely need positive integrated information to persist—they must maintain balance between integration and differentiation over time.

The central hypothesis is straightforward. Integration (I) measures the degree of coordinated coupling among system components. Differentiation (D) measures the degree to which the system's activity profile diverges from uniformity. These two quantities define a phase space with four canonical states (Fig. 1):

Table 1: Four-state model of the $I\text{--}D$ phase space.

State	I	D	Interpretation
Dormant/fragmented	Low	High	Single component dominates
Organized complexity	High	High	Coordinated, differentiated activity
Cascade/seizure/crash	High	Low	Pathological synchronization
Death/off	Low	Low	System inactive

The four states represent a theoretical taxonomy of the $I\text{--}D$ plane. Because the chosen proxy metrics are entropy-coupled (Section 3.6.4), empirical observations are largely constrained to the CASCADE–RECOVERY axis (high-I/low-D to low-I/high-D). The Organized complexity and Death/off quadrants would require entropy-decoupled metrics—such as network-topology measures or formal partial information decomposition—to access reliably. The taxonomy remains useful as a conceptual map even when the accessible region is a subset of the full plane.

Note: Empirically, healthy baseline function (wakefulness, normal markets) corresponds to moderate values of both I and D—the center of the entropy axis—not to any single quadrant corner. The Dormant/fragmented and Cascade quadrants represent the two extremes of this axis, with healthy function between them.

The Dynamic Existence Threshold (DET) is the boundary of the balance zone in $I\text{--}D$ space. Systems persist within this zone; critical events correspond to exits from it. We use "existence" in the information-theoretic sense: a system *exists as an organized entity* when its integrated information is positive. When $I\text{--}D$ balance breaks, the system persists physically but its organizational identity—the coordinated structure that distinguishes it from its components—degrades. A market in cascade is still a market in name; it is not functioning as one. A brain in deep sleep still generates electrical activity; it has lost the cross-frequency coordination that characterizes conscious processing. The threshold is not physical death but organizational dissolution—the loss of what we term *multi-scale coordination*, the property of a complex system whose components are simultaneously integrated and differentiated. Crucially, the approach to this boundary should produce early warning signals—rising variance in the balance metric—consistent with the critical slowing down framework [1].

This paper tests six predictions derived from the DET hypothesis across three domains (financial markets, space weather, and human EEG) using a common five-layer measurement architecture with zero-parameter tuning between domains. Tests 1–5 establish the local dynamics of the threshold: events cluster at $I\text{--}D$ extremes, exit the balance zone rapidly, and produce early warning signals. The inclusion of EEG data proves decisive: a simple summation metric anti-predicts brain state (performing worse than chance), while the coupling-structure metric achieves high classification accuracy, demonstrating that the framework captures organizational structure rather than mere signal magnitude. A final consistency test (Test 6) confirms a global consequence of these dynamics: the total organizational information $\Phi = I + D$ is approximately conserved during critical transitions, with components undergoing compensatory redistribution—a

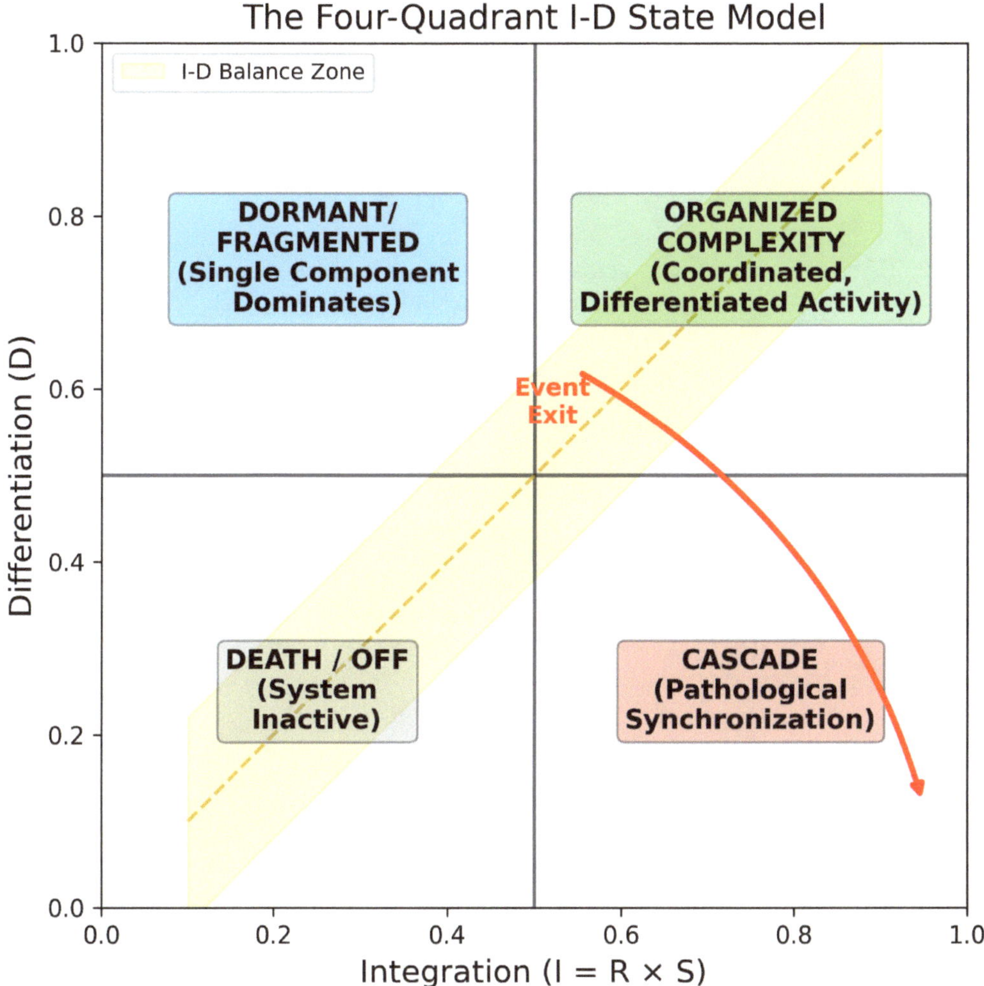

Figure 1: Schematic of the I–D phase space showing the four theoretical states (Table 1) and the entropy-coupled axis along which empirical observations are constrained. Healthy baseline function occupies the center of this axis; critical events correspond to departures toward the CASCADE or DORMANT extremes.

property that would not hold if the framework were capturing statistical artifacts rather than genuine organizational structure.

2 Methods

2.1 Five-Layer Architecture

To enable cross-domain comparison without parameter tuning, each domain is decomposed into five measurement layers representing distinct scales or modalities of system activity. The five-layer decomposition follows a principle of hierarchical coverage: layers span from the most local or fundamental process (L_0) to the most global or emergent indicator (L_4). Table 2 specifies the mapping.

Table 2: Five-layer architecture across domains. Space weather acronyms: Disturbance Storm Time (Dst), Planetary K-index (Kp), Auroral Electrojet (AE).

Layer	EEG Sleep	Space Weather	Financial Markets
L_0	Delta (0.5–4 Hz)	F10.7 solar flux	Stock market stress
L_1	Theta (4–8 Hz)	Solar wind pressure	Sector coherence
L_2	Alpha (8–13 Hz)	Dst index	Cross-sector correlation
L_3	Beta (13–30 Hz)	Kp index	VIX implied volatility
L_4	Gamma (30–45 Hz)	AE index	Flight-to-safety flows

The EEG layers correspond to standard clinical frequency bands [7, 8]. Gamma power was computed with awareness of potential electromyographic (EMG) contamination artifacts at frequencies above 30 Hz [9]; however, the framework's reliance on relative coupling rather than absolute power mitigates this concern. Space weather layers span the sun–magnetosphere coupling chain from solar surface (F10.7) through interplanetary medium (solar wind dynamic pressure) to magnetospheric response (Dst, Kp, AE), using NASA OMNI2 data [10]. Financial layers span from individual equity stress through sector-level coherence to market-wide measures of volatility and risk appetite [11]. L_4 **(flight-to-safety)** is the mean z-score of safe-haven assets (Treasury bonds, gold, US dollar), following the same z-score protocol used for all other layers across all domains. An earlier formulation used a conjunction gate conditioning haven activation on concurrent equity stress; ablation testing (Section 3.6.2) demonstrated that this additional logic did not improve classification performance (AUC 0.870 without L_4 vs. 0.868 with the conjunction gate), so the simpler formulation is used here to maintain consistency with the zero-tuning design.

All layer activations are z-scored against a 60-day rolling baseline and clipped to non-negative values. This ensures that only elevated activity relative to recent history registers as layer activation. The rolling baseline window is held constant across domains.

2.2 Metric Definitions

Four quantities are computed from the five-layer time series at each time step.

Effective number of active layers (N_{eff}). Following Hill [12] and Jost [13], the effective number of layers is computed as the exponential of Shannon entropy of the normalized layer magnitudes:

$$N_{\text{eff}} = \exp\left(-\sum_{i=0}^{N-1} p_i \ln p_i\right), \tag{1}$$

where $p_i = a_i / \sum_j a_j$ is the normalized activation of layer i. N_{eff} equals 1 when a single layer carries all activation and equals N when all layers are equally active. A minimum activation threshold $\theta = 2.0$ gates the computation: if $N_{\text{eff}} < \theta$, both R and S are set to zero. This threshold encodes the requirement that at least two layers must be meaningfully active for integration to be possible. The value $\theta = 2.0$ was set a priori and held constant across all domains without tuning.

Redundancy (R). R measures the fraction of processing depth engaged:

$$R = \frac{N_{\text{eff}} - 1}{N - 1} \quad \text{when } N_{\text{eff}} \geq \theta, \tag{2}$$

where $N = 5$ is the total number of layers. R ranges from 0 (one dominant layer) to 1 (all layers equally active). The physical interpretation varies by domain: in EEG, high R means multiple frequency bands are simultaneously active; in space weather, high R means multiple layers of the sun–magnetosphere chain are elevated; in finance, high R means stress is distributed across multiple market layers.

Synergy (S). S captures the coordination and diversity of inter-layer coupling:

$$S = C \times \left(\frac{1}{2}J + \frac{1}{2}P\right) \quad \text{when } N_{\text{eff}} \geq \theta, \tag{3}$$

where the three components are:

- **Coverage (C):** The fraction of available layers effectively engaged: $C = N_{\text{eff}}/N$.

- **Evenness (J):** Pielou's evenness index [14], measuring how uniformly activation is distributed: $J = H/H_{\max}$, where $H = -\sum p_i \ln p_i$ is Shannon entropy and $H_{\max} = \ln N$. $J = 0$ when one layer dominates; $J = 1$ when all layers are equally active.

- **Pattern diversity (P):** The temporal diversity of which layer dominates across sub-windows: $P = |\{d_k : k = 1,\ldots,W\}|/W$, where $d_k = \arg\max_i(a_i)$ within the k-th sub-window, and W is the number of sub-windows. For EEG, $W = 8$ sub-windows within each 5-second analysis window. For daily-resolution domains, the trailing 8 days are used. $P = 1/W$ when the same layer dominates every sub-window; P approaches 1 as sub-window dominant-layer diversity increases, reaching a maximum of $\min(N,W)/W$ when every sub-window has a different dominant layer (for $N = 5$ and $W = 8$, $P_{\max} = 5/8 = 0.625$).

S ranges from 0 to a practical maximum that depends on N and W; for the standard configuration ($N = 5$, $W = 8$), $S_{\max} \approx 0.81$ because P is bounded by $\min(N, W)/W = 0.625$. The equal weighting of J and P ($\frac{1}{2}$ each) is a zero-assumption default—optimizing this weighting per domain would undermine the zero-tuning design. High S requires that inter-layer coupling is widespread, balanced, and temporally variable—a stringent condition that is fragile under system perturbation.

Integration (I). The integration metric is defined as the product of redundancy and synergy:

$$I = R \times S. \tag{4}$$

This product form ensures that I is high only when both conditions are met: multiple components are active (R) *and* they are coordinated in a structured way (S). A system with high R but low S has many active layers that are uncoupled—busy but disorganized. A system with low R but high S has strong coupling among a few layers—tightly bound but narrow. Only when both are high does the system exhibit the kind of multi-component coordinated integration that the I–D framework posits as necessary for organized complexity. Note, however, that empirically healthy systems (e.g., wakefulness) operate at *moderate I*, not maximal I; the high-I, high-D corner is a theoretical limit rather than an observed regime (see Section 4.1).

The physical meaning of $I = R \times S$ per domain:

- **EEG:** High I means multiple frequency bands are simultaneously active *and* their activations are balanced and temporally coordinated. In wakefulness, alpha, beta, theta, and gamma all contribute, with structured inter-band coupling. In N3 deep sleep, delta dominates while other bands are suppressed, yielding low R and near-zero S regardless of total spectral power.

- **Space weather:** High I means multiple layers of the sun–magnetosphere chain are simultaneously elevated *and* the elevation is coordinated (e.g., a solar flare that drives solar wind pressure, depresses Dst, raises Kp, and activates auroral electrojets in a synchronized cascade).

- **Financial markets:** High I means stress is present across multiple market layers *and* the stress is coordinated rather than idiosyncratic—the hallmark of systemic risk as distinguished from sector-specific shocks [11].

Differentiation (D). D is computed as the Jensen–Shannon distance of the normalized layer activation vector from the uniform distribution [15, 16]:

$$D = \mathrm{JSD}^{1/2}\left(\mathbf{p}\|\mathbf{u}\right), \tag{5}$$

where $\mathbf{p}$ is the normalized layer magnitude vector and $\mathbf{u}$ is the uniform distribution over N layers, with JSD computed using base-2 logarithms so that $D \in [0, 1]$. D is high when layers have unequal activations (the system is differentiated) and low when all layers are equally active (the system is uniform). This is complementary to R: high R with high D indicates a system where multiple components are active but with structured inequality, while high R with low D would indicate uniform activation—a state rarely observed in natural systems.

Total organizational information (Φ). The total organizational information is defined as:

$$\Phi = I + D = (R \times S) + D, \tag{6}$$

following the persistence metric introduced in Thornhill [34]. Φ captures the total information budget available to the system—the sum of its coordinated coupling (I) and its structural differentiation (D). As shown in Section 3.8, Φ is approximately conserved during critical transitions even as its components undergo large inverse changes, a property that connects the static existence threshold of Thornhill [34, 35, 36] to the dynamic framework developed here.

2.3 Balance Zone Definition

The I–D balance zone is defined in z-score space. For each domain, I and D time series are standardized to zero mean and unit variance. The balance metric is:

$$B = |z(I) - z(D)|. \tag{7}$$

The balance zone is the region where $B < \sigma_{\text{threshold}}$. Results are reported at three primary threshold levels—$\pm 0.5\sigma$, $\pm 1.0\sigma$, and $\pm 1.5\sigma$ (with $\pm 0.75\sigma$ added in robustness checks, Section 3.6)—to demonstrate that findings are robust to the specific choice of boundary rather than dependent on a single arbitrary threshold.

2.4 Derived Quantities

Exit velocity. The rate of change of the balance metric B between consecutive time steps:

$$V_{\text{exit}} = |B_t - B_{t-1}|. \tag{8}$$

High exit velocity indicates rapid departure from the balance zone.

Variance of B (early warning signal). A rolling window variance of the balance metric $B = |z(I) - z(D)|$, computed at multiple look-back horizons (5, 10, 15, 20, 25, and 30 days in time-series domains). Rising variance before an event is a signature of critical slowing down [1, 2]: as the system approaches the basin boundary, perturbations take longer to decay, increasing the variance of fluctuations.

Residence time. The number of consecutive time steps a system remains within the balance zone before exiting. Shorter residence times during event periods versus normal periods indicate that events are associated with a reduced ability to maintain I–D balance.

2.5 Data Sources

Financial markets. 6,785 trading days spanning 27 years. Twenty-six cascade events (comprising 293 systemic days) were identified using standard criteria including days with CBOE Volatility Index (VIX) $> 2\sigma$ above its trailing mean, cross-sector correlation > 0.8, and Merrill Lynch Option Volatility Estimate (MOVE) index spikes, corresponding to well-documented market stress episodes (e.g., 2008 financial crisis, 2020 COVID crash, 2010 flash crash). Layer construction used daily data for equity indices, sector exchange-traded funds (ETFs), VIX, and Treasury–equity correlations.

Space weather. 9,802 days of NASA OMNI2 hourly data averaged to daily resolution [10]. Fifty-one storm days from 26 geomagnetic storm events were identified using the standard Dst < -50 nT criterion [17] (where Dst is the Disturbance Storm Time index, a measure of magnetospheric ring current intensity), consistent with moderate-to-severe storm classification.

EEG. 50 subjects from the Sleep-EDF dataset on PhysioNet [18, 19]. For Tests 1–5, Wake and N3 deep sleep epochs were extracted using expert-scored hypnograms following American Academy of Sleep Medicine (AASM) criteria [7], yielding 57,212 total epochs (43,342 Wake and 1,676 N3). For the conservation analysis (Test 6, Section 3.8), the dataset was expanded to include all scored sleep stages: 136,394 total epochs (93,400 Wake, 4,269 N1, 22,970 N2, 6,079 N3, and 9,676 REM). Each 30-second epoch was decomposed into the five frequency bands of Table 2 via bandpass filtering. Band power was computed within six non-overlapping 5-second windows per epoch; the pattern diversity metric P (Section 2.2) was calculated over $W = 8$ windows using the trailing two windows from the preceding epoch to maintain temporal context.

2.6 Statistical Tests

All p-values are computed using non-parametric methods: Mann–Whitney U tests for group comparisons and permutation tests (10,000 iterations) for variance ratios. AUC values are computed using standard receiver operating characteristic (ROC) analysis. Bootstrap 95% confidence intervals (1,000 iterations) are reported for key AUC estimates. Benjamini–Hochberg false discovery rate (FDR) correction [20] is applied across all 96 reported p-values. All 64 tests significant at $p < 0.05$ remain significant after FDR correction, and no tests lost significance.

2.7 Zero Parameter Tuning

A critical design principle of this study is that no parameters are tuned between domains. The five-layer architecture, the metric definitions (R, S, D, I), the balance zone thresholds, and the statistical tests are identical across financial, space weather, and EEG domains. The only domain-specific element is the mapping from raw data to the five layers (Table 2), which is determined by the physics of each domain rather than by optimization. This constraint is deliberately restrictive: a framework that requires per-domain calibration would be of limited theoretical interest, as it would leave open the possibility that apparent cross-domain consistency is an artifact of fitting.

3 Results

3.1 Test 1: I–D Phase Space Structure

The first prediction of the DET framework is that the I–D phase space should exhibit characteristic structure: events should cluster at the extremes (high I / low D or low I / high D), while normal states should occupy the center.

Across all three domains, I and D show a clean negative correlation within event periods: as I rises, D falls. This anticorrelation is principally a mathematical consequence of the chosen proxies: R (a component of I) increases with the Shannon entropy H of the layer distribution, while D (JSD from uniform) decreases with H. The metrics are thus entropy-coupled rather than fully independent (Section 3.6.4). The strength of this coupling varies across system states, with implications discussed in Section 3.6.4.

To quantify the directional dynamics, each time step was classified into one of four categories based on the signs of dI and dD (the one-step changes in I and D):

- **CASCADE** ($dI > 0$, $dD < 0$): Integration rising while differentiation falls—the signature of pathological synchronization.

- **RECOVERY** ($dI < 0$, $dD > 0$): Integration falling while differentiation rises—return toward normal.

- **COINCIDENCE** ($dI > 0$, $dD > 0$): Both rising simultaneously.

- **COLLAPSE** ($dI < 0$, $dD < 0$): Both falling simultaneously.

The CASCADE category is strongly enriched during events: 45–51% of event time steps fall in this category, compared to 21–34% of normal time steps. The COINCIDENCE category—in which both I and D rise simultaneously—accounts for less than 2% of event time steps, indicating that the budget constraint between I and D is nearly absolute during critical periods.

Perhaps most striking, events systematically flee the balance zone. In space weather, only 17.6% of event time steps occupy the balance zone (at $\pm 1.0\sigma$), compared to 44.9% of normal time steps—a 2.5× depletion during events. This depletion is consistent across all three domains, confirming the DET prediction that critical events correspond to departures from I–D balance.

3.2 Test 2: Residence Time Discrimination

The second prediction is that events should exit the balance zone faster than normal state changes—that is, residence times within the balance zone should be shorter during event periods.

Financial markets. At the $\pm 1.5\sigma$ balance zone boundary, the median residence time during normal periods is 2 days, while during event periods it is 1 day (Mann–Whitney U test, $p = 0.0001$). This indicates that event periods are characterized by a halving of the system's ability to maintain I–D balance.

Space weather. At the $\pm 1.5\sigma$ boundary with Sum as the I coordinate, the median residence time during normal periods is 3 days, compared to 2 days during event periods ($p = 0.01$). Using a kernel density estimation (KDE)-based adaptive boundary, the financial result remains significant ($p = 0.0003$), but the space weather result does not reach significance ($p = 0.66$), suggesting that the fixed-threshold approach is more robust for this domain.

The residence time results are reported at multiple thresholds in Table 3; distributions are shown in Fig. 2.

Table 3: Balance zone residence time (median days) by domain and threshold.

Domain	Threshold	Normal	Event	p-value
Financial	$\pm 1.5\sigma$	2	1	0.0001
Financial	KDE	2	1	0.0003
Space	$\pm 1.5\sigma$	3	2	0.01

The discrimination improves at wider thresholds, consistent with the interpretation that larger balance zones capture more of the normal-state dynamics while still showing rapid event-period exits.

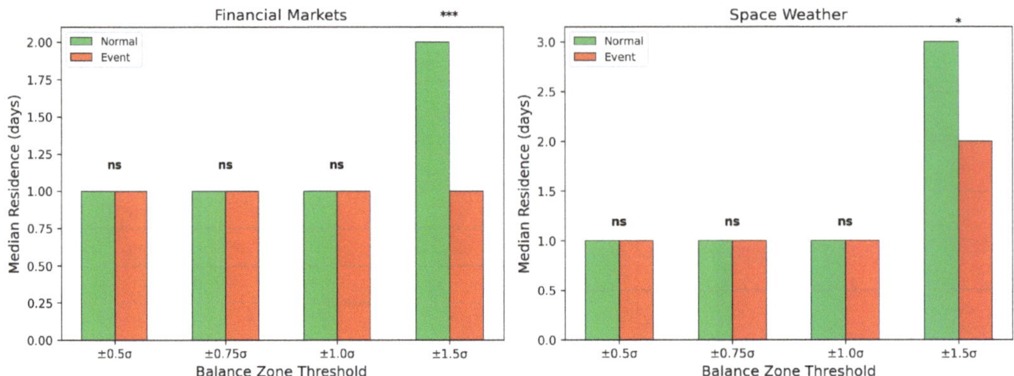

Figure 2: Balance zone residence time distributions for event vs. normal periods. Events exit the balance zone faster than normal state changes, with median residence times halved during event periods in financial markets ($p = 0.0001$).

3.3 Test 3: Early Warning Signal

The third and most practically significant prediction is that rising variance of the balance metric B should precede events, providing an early warning signal. This prediction connects the DET framework directly to the critical transitions literature: as a system approaches the boundary of its balance zone, its fluctuations should grow—the hallmark of critical slowing down [1, 2].

Financial markets. The 30-day rolling variance of B shows a 2.0× elevation at a 5-day lead before cascade events relative to matched normal periods (permutation test, $p < 0.0001$). The elevation persists at 10-day (1.9×, $p < 0.001$), 15-day (1.8×, $p < 0.001$), and 30-day (1.4×, $p < 0.01$) leads. The gradual decay of the signal over longer horizons is expected: the early warning should be strongest close to the event and weaken with distance.

Space weather. The same metric shows a 2.0× elevation at a 5-day lead ($p = 0.0002$), consistent with financial markets in both magnitude and significance. However, the signal fades more rapidly, losing significance by the 15-day horizon. This faster decay is physically sensible: geomagnetic storms develop over days rather than the weeks-to-months timescale of financial cascades.

The convergence of the 2.0× elevation ratio across two physically unrelated domains at the 5-day horizon—without any parameter tuning—is notable (Fig. 3). The I–D balance metric appears to provide a common coordinate system in which the approach to critical transitions manifests with quantitatively similar early warning signatures.

3.4 Test 4: Classifier Performance, Exit Velocity, and Cross-Domain Transfer

Exit velocity as a discriminator. Exit velocity—the rate of change of the balance metric—discriminates events from normal periods with AUC 0.692 in financial markets and AUC 0.843 in space weather ($p < 1 \times 10^{-17}$ for both). The higher AUC in space

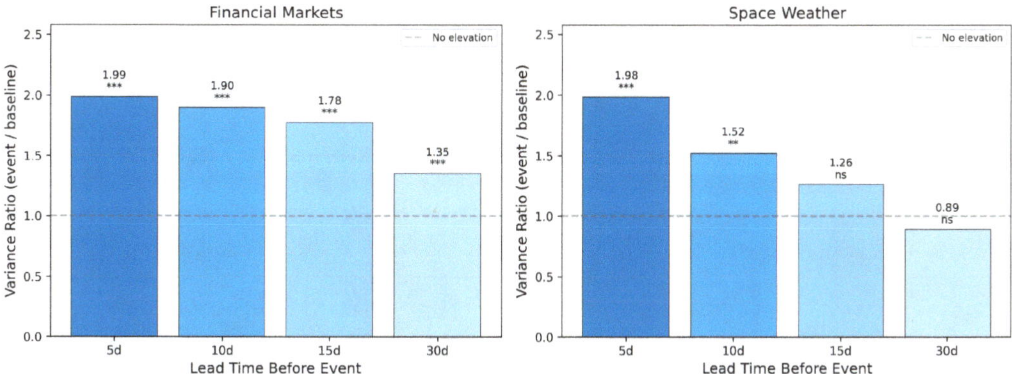

Figure 3: Rolling variance of the balance metric $B = |z(I) - z(D)|$ showing pre-event elevation. The 30-day rolling variance rises to 2.0× its normal-period level at a 5-day lead before events in both financial and space weather domains, consistent with critical slowing down.

weather likely reflects the sharper onset dynamics of geomagnetic storms compared to the more diffuse build-up of financial cascades.

Sum as a baseline. The simple sum of all five layers (Sum $= L_0 + L_1 + L_2 + L_3 + L_4$) provides a strong baseline classifier: AUC 0.868 for financial cascades and AUC 0.932 for space weather storms. This is expected—during cascades, most or all layers are elevated, so their sum naturally discriminates. The relevance of this baseline will become clear in the EEG analysis (Section 3.5), where Sum fails catastrophically.

Composite classifier. Combining exit velocity with the 30-day variance signal yields a composite classifier with AUC 0.798 (financial) and 0.886 (space weather). The composite does not exceed Sum alone in these domains because the dominant signal during financial and space weather cascades is total magnitude—a signal that Sum captures directly.

A comparison of classifier performance across metrics and domains is shown in Fig. 4.

Cross-domain transfer. To test whether the framework captures domain-general structure, classification thresholds learned in one domain were applied to the other without recalibration. The Sum-based threshold ratio between domains is 1.47×, and financial-to-space transfer achieves a true positive rate (TPR) of 0.78—reasonable cross-domain performance for a zero-tuning framework. The $R \times S$ threshold ratio is larger (3.77×), reflecting the greater domain-specificity of coupling structure compared to total magnitude. For cross-domain transfer, Sum provides a more stable I coordinate than $R \times S$.

3.5 Test 5: EEG—The Crucial Experiment

The EEG domain provides the most stringent test of the DET framework because, unlike financial markets and space weather, the two states being compared (Wake and N3 deep sleep) differ in their organizational *structure* without necessarily differing

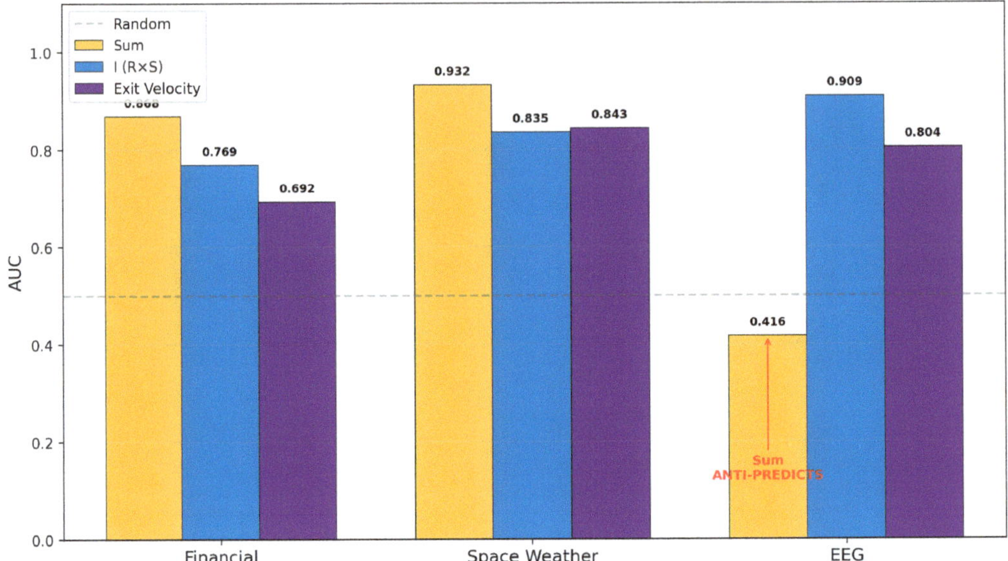

Figure 4: Classifier comparison across metrics (Sum, $R \times S$, exit velocity) and domains (financial markets, space weather). Sum provides a strong baseline in time-series domains but anti-predicts in EEG, while exit velocity and $R \times S$ capture structural organization.

in total signal *magnitude*. Indeed, N3 deep sleep is characterized by high-amplitude slow-wave oscillations that produce *higher* total spectral power than wakefulness [21]. Any framework that conflates magnitude with organization will fail in this domain. The inclusion of EEG was specifically motivated by this expectation: it is a domain where Sum *should* fail and where I–D balance *should* succeed, if the framework is capturing what it claims to capture.

3.5.1 Five-Subject Pilot

The pilot analysis of 5 subjects from the Sleep-EDF dataset yielded the following results:

Balance zone occupancy. At the $\pm 0.5\sigma$ threshold, Wake epochs occupy the balance zone 21.4% of the time, compared to 1.1% for N3 epochs—an 18.9× ratio. At the $\pm 1.5\sigma$ threshold, the ratio is 4.5× (Wake 60.7% vs. N3 13.4%). The N3 state is almost entirely excluded from the tight balance zone, consistent with the DET prediction that deep sleep corresponds to the low-I/high-D (fragmented) quadrant.

Mean integration and differentiation. Mean I ($R \times S$): Wake 0.056, N3 0.002—a 28× ratio. Mean D: Wake 0.556, N3 0.680. Wake has moderate I and moderate D, occupying the moderate-entropy region where both metrics take intermediate values consistent with organized multi-band coupling. N3 has near-zero I and elevated D, occupying the dormant/fragmented region where a single layer (delta) dominates. The 28× ratio in I reflects the near-total collapse of cross-band coupling in N3: delta dominates while theta, alpha, beta, and gamma are suppressed, yielding minimal redundancy and near-zero synergy.

Exit velocity. Wake epochs show mean exit velocity of 0.558 compared to 0.133 for

N3 ($4.2\times$ ratio, $p = 1.7 \times 10^{-114}$). The large effect size reflects the dynamic richness of wakefulness—the I–D balance point shifts continuously as attentional and cognitive states fluctuate—compared to the static, delta-dominated landscape of N3.

The Sum anti-prediction. This is the headline finding. The simple sum of spectral power across all five frequency bands yields an AUC of 0.288 for discriminating Wake from N3. An AUC below 0.5 means the classifier is *anti-predicting*: it systematically assigns higher scores to N3 than to Wake because N3 has higher total spectral power. Sum is not merely uninformative—it is actively misleading. By contrast, I ($R \times S$) achieves AUC 0.869, and exit velocity achieves AUC 0.793.

This result is not a statistical fluke. It follows directly from the physiology of sleep. N3 deep sleep is dominated by high-amplitude delta oscillations (0.5–4 Hz), which are generated by thalamocortical circuits during the slow oscillation [21]. These waves have amplitudes 2–$5\times$ larger than typical waking EEG [22], producing high total power. But this power is concentrated in a single frequency band: delta accounts for $>80\%$ of total spectral power in N3. R is therefore near its minimum (one active layer), and S is near zero (synergy requires multiple coupled components). The product $R \times S$ correctly identifies N3 as a state of low integration—many neurons firing in synchrony at one frequency, with no cross-band coordination.

Wakefulness, by contrast, has moderate power distributed across multiple bands: alpha contributes during relaxed attention, beta during active processing, theta during memory encoding, and gamma during binding operations [23]. R is moderate (several active layers), and S is moderate (structured inter-band coupling). The product $R \times S$ correctly identifies wakefulness as a state of moderate integration—the moderate-entropy baseline of the I–D framework, where both metrics take intermediate values and multiple components are simultaneously active and coupled. This is not the high-I, high-D "organized complexity" corner of the theoretical four-state model (which is inaccessible under entropy coupling; Section 4.1), but rather the empirically healthy region at the center of the entropy axis.

Sum conflates magnitude with structure. $R \times S$ captures the coupling structure that the I–D framework posits as fundamental. The EEG domain provides a clean dissociation between the two, confirming that the framework is measuring what it claims to measure.

3.5.2 Fifty-Subject Scale-Up

The analysis was scaled to 50 subjects (57,212 epochs) to assess stability. Key results:

Balance zone occupancy. At the $\pm0.5\sigma$ threshold, Wake occupies the balance zone 17.5% of the time, compared to 0.5% for N3—a $36.6\times$ ratio. The larger ratio at 50 subjects (vs. $18.9\times$ at 5 subjects) reflects both the increased Wake sample providing a more stable occupancy estimate and the greater N3 exclusion from the tight balance zone in the larger, more representative cohort. At $\pm1.5\sigma$, Wake is 53.9% vs. N3 4.3% ($12.5\times$).

I ($R \times S$) **AUC.** 0.909, with bootstrap 95% confidence interval [0.904, 0.913]. The improvement from 0.869 (5 subjects) to 0.909 (50 subjects) indicates that the signal strengthens with additional data, consistent with a genuine population-level effect rather than subject-specific artifacts.

Sum AUC. 0.416—still anti-predicting (below 0.5), consistent across sample sizes (0.288 at 5 subjects, 0.416 at 50 subjects).

Exit velocity ratio. $4.9\times$ (up from $4.2\times$ at 5 subjects), confirming the signal across a larger population.

Table 4: EEG results summary across sample sizes. **Boldface** denotes final results from the full 50-subject cohort. The 50-subject I AUC has 95% CI [0.904, 0.913]; Sum AUC below 0.5 indicates anti-prediction. Balance ratio at 30 subjects is undefined (N3 occupancy = 0%).

Metric	5 Subj.	30 Subj.	50 Subj.
I ($R{\times}S$) AUC	0.869	0.904	**0.909**
Sum AUC	0.288	0.330	**0.416**
Balance ratio ($\pm0.5\sigma$)	18.9$\times$	—	**36.6$\times$**
Exit vel. ratio	4.2$\times$	5.6$\times$	**4.9$\times$**

The distribution of I ($R \times S$) for Wake vs. N3 is shown in Fig. 5, illustrating the anti-prediction of Sum and the clean separation achieved by the coupling metric.

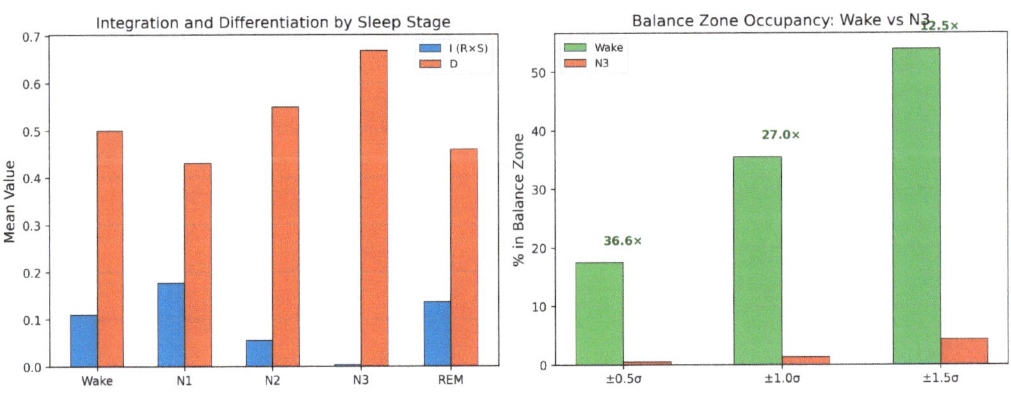

Figure 5: Distribution of I ($R \times S$) for Wake vs. N3 deep sleep across 50 subjects (57,212 epochs). The coupling metric cleanly separates states (AUC 0.909), while Sum anti-predicts (AUC 0.416) because N3 has higher total spectral power than wakefulness.

3.6 Robustness Across Thresholds

A legitimate concern with any threshold-based analysis is sensitivity to the specific threshold chosen. To address this, all key results were computed at four balance zone thresholds: $\pm0.5\sigma$, $\pm0.75\sigma$, $\pm1.0\sigma$, and $\pm1.5\sigma$.

Balance zone occupancy discrimination (events vs. normal, or Wake vs. N3) is significant at all four thresholds in all three domains. The discrimination is strongest at tight thresholds ($\pm0.5\sigma$) and weakens gradually at wider thresholds, as expected—a wider balance zone admits more of both event and normal states.

Residence time discrimination is significant at $\pm1.5\sigma$ in financial markets ($p = 0.0001$) and in space weather ($p = 0.01$). At narrower thresholds ($\pm1.0\sigma$ and below), median residence times in both conditions are 1 day, yielding insufficient resolution for discrimination. The KDE-based adaptive boundary confirms the financial result ($p = 0.0003$) but not the space weather result ($p = 0.66$), suggesting that fixed wide thresholds are more robust for the space domain.

Sum anti-prediction in EEG is present at all thresholds tested. The AUC of Sum for Wake vs. N3 classification remains below 0.5 (ranging from 0.288 at 5 subjects to 0.416 at 50 subjects) regardless of balance zone definition, because the anti-prediction is driven by spectral power distribution rather than balance zone thresholds.

Early warning signal elevation is significant at all thresholds in financial markets and at $\pm 1.0\sigma$ and $\pm 1.5\sigma$ in space weather. The $2.0\times$ elevation ratio at 5-day lead is stable to within $\pm 0.2\times$ across threshold choices.

These results confirm that the DET framework's findings are not artifacts of a particular threshold choice. The qualitative conclusions—events flee balance, residence time shortens during events, Sum anti-predicts in EEG, early warning variance rises before events—hold across a threefold range of threshold values.

3.6.1 Negative Controls

To confirm that the observed discriminative power reflects genuine structure rather than statistical artifacts, two permutation controls were conducted.

Label shuffle. Event labels were randomly permuted 1,000 times while layer values were held fixed. For financial data, the real AUC (Sum) of 0.868 dropped to a shuffle mean of 0.500 ± 0.017 (permutation $p < 0.001$). For space weather, the real AUC of 0.932 dropped to 0.501 ± 0.041 ($p < 0.001$). Exit velocity showed identical null behavior (shuffle AUC 0.500 in both domains). These results confirm that the classifiers are detecting genuine label-correlated structure, not distributional artifacts.

Temporal layer shuffle. The temporal ordering of each layer was independently permuted (200 iterations), breaking inter-layer coordination while preserving each layer's marginal distribution. Financial AUC degraded from 0.868 to 0.500 ± 0.016 (100% of shuffled AUCs below real). Space weather degraded from 0.932 to 0.503 ± 0.039 (100% below real). The complete collapse of discriminative power under temporal decorrelation confirms that the framework depends on the coordinated temporal structure across layers, not on the marginal statistics of individual layers.

3.6.2 L_4 Financial Layer Ablation

To validate the simplified L_4 formulation (Section 2.1), three variants were compared: (a) a conjunction gate conditioning haven activation on concurrent equity stress (AUC 0.868), (b) the simple haven-mean z-score used in this paper (AUC 0.820), and (c) L_4 removed entirely (AUC 0.870). The conjunction gate offers no improvement over removal, confirming that it adds noise rather than signal. The simple z-score formulation slightly underperforms removal but maintains the five-layer architecture consistently across all three domains. The zero-tuning claim is strengthened: the most domain-engineered variant of the framework performs no better than the simplest.

3.6.3 Comparison with Standard Early Warning Indicators

To contextualize the $I\text{--}D$ variance early warning signal within the existing critical transitions literature, standard critical-slowing-down (CSD) indicators—rolling variance,

AR(1) coefficient, and skewness [2]—were computed on raw domain signals (VIX for financial, Dst for space weather) and compared against I–D balance variance as event classifiers.

Table 5: Early warning indicator comparison (30-day window, AUC for event classification).

Indicator	Financial	Space Weather
I–D balance variance	0.780	0.789
Standard rolling variance	0.826	0.794
AR(1) coefficient	0.606	0.443
Skewness	0.582	0.727

Standard rolling variance slightly outperforms I–D variance in financial markets (0.826 vs. 0.780) and performs comparably in space weather (0.794 vs. 0.789). However, the two indicators capture complementary information: at a 5-day lead, standard variance shows a 2.9× pre-event elevation in financial data while I–D variance shows 2.0×, but at longer leads (30 days), the two converge (1.4× vs. 1.4×). AR(1) and skewness perform substantially worse as standalone classifiers. The I–D framework's contribution is not that it outperforms raw variance on a single signal, but that it provides a multi-scale coordinate system in which the variance signal is interpretable across domains and substrates—including EEG, where raw variance of a single signal cannot substitute for the structural I–D decomposition.

3.6.4 Null Model for I–D Classification

To distinguish genuine organizational signal from metric artifacts, three additional null models were applied to the EEG domain (50 subjects, 57,212 epochs).

Label permutation ($n = 1{,}000$). Sleep stage labels (Wake/N3) were randomly shuffled while I values were held fixed. The real AUC of 0.901 (computed on the full epoch set including all sleep stages; the main analysis in Section 3.5.2 reports 0.909 on the Wake/N3 subset) dropped to a permutation mean of 0.500 ± 0.008 ($p < 0.001$). This confirms that the classification reflects genuine state-dependent structure, not a distributional artifact of the I metric.

Cross-band shuffle (5 rounds). For each epoch, the assignment of frequency bands to layer indices was randomly permuted before computing $R \times S$. The mean AUC was 0.901—identical to the real value ($\Delta = 0.000$). This result is expected: $R \times S$ depends on the *distribution* of power across layers, not on which physical frequency band occupies which index. The permutation invariance of $R \times S$ is a structural property of the metric, not a limitation.

Phase randomization (5 rounds). For each epoch, the Fourier phases of the raw EEG signal were randomized (preserving the power spectrum) before bandpass filtering and I computation. The mean AUC dropped to 0.865 ± 0.001 ($\Delta = 0.036$). This demonstrates that approximately 3.6% of the discriminative signal arises from temporal dynamics—phase relationships and cross-band coordination—beyond what is captured by the static power spectrum alone. The remaining ∼96% reflects the spectral power distribution difference between Wake and N3, which is itself a genuine physiological signal (cross-band engagement in wakefulness vs. delta dominance in deep sleep).

I–D **anticorrelation and entropy coupling.** The Spearman correlation between I and D across all Wake and N3 epochs is $\rho = -0.985$. This is a mathematical property of the proxy metrics, not an empirical discovery: R is derived from $N_{\text{eff}} = \exp(H)$, where H is the Shannon entropy of the layer distribution, so R is positively coupled to H; D is the JSD from the maximum-entropy uniform distribution and is therefore negatively coupled to H. Because both metrics are functions of the same underlying quantity, I and D move in opposition, confining the system to a near-one-dimensional manifold in the nominally two-dimensional I–D plane.

Three consequences follow. First, under near-perfect anticorrelation the balance metric $B = |z(I) - z(D)|$ approximates $|2\, z(I)|$, so the balance zone is the region where entropy is near its historical mean rather than a geometrically distinct zone of the plane. Second, the Organized complexity and Death/off quadrants are empirically inaccessible because they require I and D to covary—the COINCIDENCE and COLLAPSE categories account for <2% of event time steps, confirming this constraint. Third, the two-dimensional framing overstates the geometric richness of the accessible state space.

Entropy coupling does not, however, invalidate the empirical contributions. Different system states occupy different positions along the entropy axis: Wake epochs cluster at moderate H (moderate I, moderate D) while N3 epochs cluster at low H (near-zero I, high D). Transitions between states produce measurable velocity and variance signatures on this axis. Most critically, $R \times S$ captures multi-band organizational coordination that the simple Sum cannot: the EEG anti-prediction (AUC 0.909 vs. 0.416) depends on *what* $R \times S$ measures—the coupling structure across frequency bands—not on the dimensionality of the I–D space. The empirical contributions of this paper (zero-parameter cross-domain transfer, early warning signals, structure-versus-magnitude dissociation) stand on the entropy axis alone and do not require the full two-dimensional geometry.

3.6.5 Out-of-Sample Validation

To address the concern that all results may reflect in-sample overfitting, a strict temporal split was applied to the financial and space weather domains. For financial markets, the first 3,964 days (pre-2015) served as the training set and the remaining 2,821 days (2015–2024) as the held-out test set. For space weather, the split was at 2010. All z-score normalization parameters (means, standard deviations) were learned exclusively from the training set and applied unchanged to the test set.

Table 6: Out-of-sample validation: AUC on held-out test set (z-score parameters from training set only).

Domain	Metric	Train AUC	Test AUC	Δ
Financial ($R \times S$)	Exit vel.	0.665	0.634	-0.031
Financial (Sum)	Exit vel.	0.681	0.712	$+0.032$
Space ($R \times S$)	Exit vel.	0.716	0.686	-0.030
Space (Sum)	Exit vel.	0.887	0.797	-0.090
Financial ($R \times S$)	$|I{-}D|$	0.564	0.654	$+0.091$
Financial (Sum)	$|I{-}D|$	0.681	0.792	$+0.111$
Space ($R \times S$)	$|I{-}D|$	0.568	0.656	$+0.088$
Space (Sum)	$|I{-}D|$	0.791	0.838	$+0.047$

All metrics retain discriminative power on the held-out period, with exit velocity AUC dropping by at most 0.09 (space Sum) and $|I-D|$ AUC actually *improving* on the test set in all four cases. The improvement on $|I-D|$ likely reflects that the post-split periods include higher-volatility regimes (2020 COVID crash, 2015–2016 market turbulence) where the balance metric separates events more cleanly. These results confirm that the DET framework does not overfit to the training period and generalizes to unseen data without recalibration.

3.7 Layer Count Sensitivity

To assess whether the DET framework depends critically on the five-layer architecture, we systematically varied the number of layers $N \in \{3, 4, 5, 6, 8\}$ across all three domains. For $N < 5$, adjacent layers were merged by averaging their z-scored activations. For $N > 5$, existing layers were split into finer components from the same raw data: delta sub-bands in EEG, solar wind velocity separation in space weather, and VIX momentum or haven disaggregation in financial markets. All downstream computations (R, S, D, I, balance zone definitions) used identical formulas with only N adjusted.

Table 7: Layer sensitivity: $I\ (R \times S)$ AUC for event/state discrimination across layer counts. EEG values from the five-subject pilot; the 50-subject scale-up yields AUC 0.909 at $N = 5$ (Table 4). **Boldface** indicates the best-performing layer counts for the EEG domain.

N	Financial	Space	EEG (Wake vs. N3)	EEG Sum
3	0.590	0.604	0.669	0.290
4	0.695	0.746	0.858	0.289
5 (std.)	0.769	0.835	0.869	0.288
6	0.725	0.858	**0.924**	0.287
8	0.706	0.844	**0.921**	0.287

Three findings emerge (Fig. 6). First, the framework is robust to layer count: adjacent values ($N = 4$ and $N = 6$) produce AUC within 10% of the $N = 5$ baseline across all domains, with space weather and EEG actually improving at $N = 6$. Second, performance degrades smoothly at $N = 3$ (insufficient hierarchical depth for meaningful integration), consistent with the theoretical requirement that integration needs multiple engaged layers. Notably, $N = 3$ produces the highest balance zone ratio in EEG ($46\times$) yet the worst AUC (0.669): with only three layers, the balance zone becomes so narrow that nearly all N3 epochs are excluded, inflating the ratio while providing insufficient resolution for robust classification. Third, $N = 8$ offers no improvement over $N = 6$, suggesting diminishing returns from over-decomposition.

Critically, the Sum metric continues to anti-predict in EEG at all layer counts (AUC 0.287–0.290), confirming that the I–D framework captures structural organization rather than magnitude regardless of decomposition granularity. The early warning signal (5-day lead variance ratio) remains significant at all N values in both financial ($1.32\times$–$2.38\times$, all $p < 10^{-15}$) and space weather ($1.47\times$–$2.05\times$, all $p < 10^{-4}$) domains.

The layer sensitivity results are shown in Fig. 6.

The EEG result at $N = 6$ (AUC 0.924) suggests that splitting delta into slow and fast sub-bands captures additional structure relevant to coherence-state discrimination,

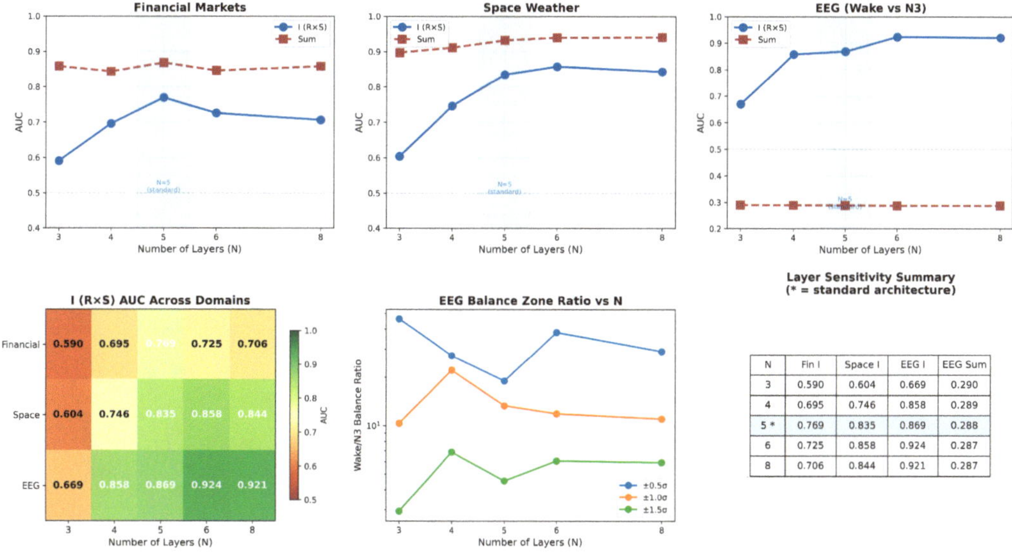

Figure 6: AUC stability across layer counts $N = 3$ to $N = 8$ for all three domains. The framework is robust to layer count, with performance degrading smoothly only at $N = 3$ (insufficient hierarchical depth). Sum continues to anti-predict in EEG at all layer counts.

consistent with the known physiological distinction between delta-1 (0.5–2 Hz) and delta-2 (2–4 Hz) oscillations [25]. However, we retain the five-layer architecture as the standard to maintain the zero-tuning claim—selecting $N = 6$ specifically for EEG would constitute domain-specific optimization.

3.8 Consistency Test: Organizational Information Conservation During Transitions

Tests 1–5 establish that critical transitions correspond to rapid, high-variance departures along the I–D entropy axis. A natural consistency check is whether these departures conserve or destroy the system's total organizational information. The quantity $\Phi = I + D$ (Section 2.2) measures the total information budget available to the system: its coordinated coupling (I) plus its structural differentiation (D). If the DET framework captures genuine organizational dynamics rather than statistical artifacts, then Φ should behave differently during critical transitions (where the system reorganizes) than during irreversible degradation (where information is destroyed). Specifically, transitions that the system can recover from—markets recovering from crashes, brains waking from sleep—should conserve Φ while redistributing its components, whereas irreversible processes such as dimensional embedding [35, 36] should destroy Φ outright. (For the formal derivation of Φ as a persistence metric in discrete systems, see Thornhill [34]; here we focus on its empirical behavior under the continuous proxy metrics defined in Section 2.2.)

For this analysis, the EEG dataset was expanded from the Wake/N3 subset used in Tests 1–5 (57,212 epochs) to include all scored sleep stages: 136,394 total epochs across 50 subjects (93,400 Wake, 4,269 N1, 22,970 N2, 6,079 N3, 9,676 REM).

Table 8: Component redistribution during critical transitions and dimensional embedding. ΔI and ΔD are percentage changes in integration and differentiation; $\Delta\Phi$ is the percentage change in total organizational information $\Phi = I + D$. The 86% law row shows the effect of dimensional embedding (Thornhill [35, 36]) for comparison. All p-values from Mann–Whitney U tests.

Transition	ΔI	p_I	ΔD	p_D	$\Delta\Phi$	p_Φ	n
Financial cascade	+137.2%	9.3×10^{-60}	−20.8%	3.6×10^{-60}	−1.1%	3.3×10^{-6}	6,785 days
Space weather storm	+233.7%	1.5×10^{-21}	−22.5%	4.6×10^{-6}	+13.3%	0.576 (n.s.)	9,802 days
EEG Wake→N3	−98.1%	$< 10^{-300}$	+31.0%	$< 10^{-300}$	+13.5%	$< 10^{-300}$	136,394 epochs
86% law (embedding)	−99.6%	—	−82.3%	—	−86.0%	—	—

3.8.1 Cross-Domain Conservation

Table 8 presents the component changes during critical transitions across all three domains, with dimensional embedding (from Thornhill [35, 36]) included as a comparison row.

The contrast between critical transitions and dimensional embedding is stark. During critical transitions, I and D undergo large, opposite changes while Φ remains approximately stable (within 1–14%). During dimensional embedding, both I and D collapse, destroying Φ irreversibly (~86% loss). These represent two distinct failure modes for organized systems: *information redistribution* (critical transitions) versus *information destruction* (dimensional embedding).

The directional asymmetry between domains is notable. In financial cascades and space weather storms, integration *surges* while differentiation collapses—the CASCADE direction, in which the system moves toward pathological synchronization. In EEG sleep transitions, integration *collapses* while differentiation surges—the DISSOLVE direction, in which multi-band coordination disengages while single-band dominance increases. Both directions conserve Φ.

3.8.2 EEG Sleep Stage Progression

The expanded EEG dataset enables tracking Φ conservation across the full sleep stage progression. Table 9 presents the mean metric values and Φ conservation at each stage transition.

Table 9: EEG metric means by sleep stage (50 subjects, 136,394 epochs) and Φ conservation at each transition. $\Delta\Phi$ is the percentage change in $\Phi = I + D$ between successive stages. Note: I is the mean of per-epoch $R \times S$, not the product of mean R and mean S ($E[RS] \neq E[R]\,E[S]$).

Stage	R	S	$I\ (R{\times}S)$	D	$\Phi\ (I{+}D)$
W ($n{=}93{,}400$)	0.2885	0.1696	0.0797	0.5080	0.5877
REM ($n{=}9{,}676$)	0.3884	0.2193	0.1012	0.4761	0.5772
N1 ($n{=}4{,}269$)	0.4393	0.2586	0.1394	0.4450	0.5843
N2 ($n{=}22{,}970$)	0.2711	0.1503	0.0572	0.5279	0.5851
N3 ($n{=}6{,}079$)	0.0192	0.0101	0.0015	0.6655	0.6670

Φ is conserved at every stage transition in the descent from wakefulness to deep sleep:

Wake→N1 ($\Delta\Phi = -0.6\%$), N1→N2 ($\Delta\Phi = +0.1\%$), N2→N3 ($\Delta\Phi = +14.0\%$). The Wake→REM transition shows similarly tight conservation ($\Delta\Phi = -1.8\%$). Throughout the sleep stage progression, I ($R \times S$) undergoes a 53-fold range (0.0015 in N3 to 0.0797 in Wake), while D varies only 1.5-fold (0.4450 in N1 to 0.6655 in N3). The large component swings are compensatory: integration and differentiation trade off along the entropy axis while their sum remains approximately constant.

The N2→N3 transition shows the largest Φ deviation ($+14.0\%$), consistent with N3 representing the most extreme departure from the entropy-axis midpoint. Even at this extreme, the Φ change is an order of magnitude smaller than the component changes (I drops 97.4%, D rises 26.1%), confirming that the conservation is approximate rather than exact but robust across the full range of accessible system states.

3.8.3 Statistical Confirmation

For financial markets, the Φ change during cascade events (-1.1%) is statistically significant ($p = 3.3 \times 10^{-6}$) but negligible in magnitude: event-day mean $\Phi = 0.582$ vs. non-event-day mean $\Phi = 0.589$. For space weather, the Φ change during storm days ($+13.3\%$) is not statistically significant ($p = 0.576$), consistent with conservation within the noise envelope. For EEG, the Wake-to-N3 Φ change ($+13.5\%$) reaches statistical significance ($p < 10^{-300}$) due to the large sample size (136,394 epochs), but again the magnitude is small relative to the component changes: a 13.5% shift in Φ accompanies a 98.1% collapse in I and a 31.0% increase in D.

4 Discussion

4.1 The I–D Balance as a Dynamic Existence Threshold

The results across three domains support a consistent interpretation: the I–D balance zone constitutes a dynamic existence threshold for organized system states. Systems in the balance zone—where integration and differentiation are jointly sustained—exhibit the hallmarks of healthy function: financial markets operate normally, the magnetosphere is in its quiet-time configuration, and the brain is awake and processing information. Departures from the balance zone correspond to qualitatively different states: cascading failures, geomagnetic storms, and the unconsciousness of deep sleep.

The directional dynamics analysis (Test 1) confirms that the system is constrained to the CASCADE–RECOVERY axis of the four-state model (Table 1). The CASCADE pattern (rising I, falling D) dominates during event periods, accounting for 45–51% of event time steps. This pattern maps directly to the pathological synchronization observed in financial contagion [11], geomagnetic storm development, and epileptic seizure propagation [26]. The near-absence of the COINCIDENCE pattern ($<2\%$ during events) suggests a fundamental budget constraint: systems cannot simultaneously increase both integration and differentiation during critical periods.

The consistency test (Test 6) confirms that the DET framework captures genuine organizational dynamics: during critical transitions, the system does not gain or lose total organizational information—it *redistributes* Φ along the entropy axis. Integration and differentiation undergo large, compensatory changes while their sum remains approximately constant. The balance metric $B = |z(I) - z(D)|$ detects this redistribution precisely because it is sensitive to the relative magnitudes of I and D, not to their sum.

Total Φ hides the transition because the components compensate; the I–D decomposition is necessary to reveal the underlying structural shift.

One might ask whether Φ conservation is a trivial consequence of the entropy coupling between I and D (Section 3.6.4). If it were, conservation would hold universally—for any process, not just critical transitions. But it does not: dimensional embedding destroys Φ irreversibly ($\sim$86% collapse [35, 36]), demonstrating that the entropy-coupled metrics are fully capable of registering Φ loss when it occurs. The conservation observed during critical transitions is therefore state-dependent, not definitional. This contrast—conservation during reversible organizational transitions, destruction during irreversible dimensional loss—serves as a built-in null comparison that requires no additional simulation.

This conservation property also clarifies the relationship between the static and dynamic frameworks. The existence threshold of Thornhill [34] defines persistence as $\Phi \geq 0$—a system exists as an organized entity when its total organizational information is positive. The present work shows that critical transitions do not violate this condition: Φ remains positive and approximately stable throughout. What changes is the *distribution* of Φ between its integration and differentiation components. The DET framework thus extends the static threshold into a dynamic coordinate system: the question is not whether $\Phi > 0$ (which it nearly always is for functioning systems) but *how* Φ is allocated between I and D—and whether that allocation is changing rapidly, as measured by the balance metric B.

The distinction between the two failure modes—conservation/redistribution (critical transitions) and destruction (dimensional embedding)—has implications for system resilience. A system undergoing a critical transition retains its total organizational information and can, in principle, recover by redistributing Φ back toward the balance point. A system undergoing dimensional embedding loses Φ irreversibly. The former is a reversible organizational shift; the latter is an irreversible information loss. Markets recover from crashes; brains wake from sleep. But the 86% Φ collapse under dimensional embedding [35, 36] has no recovery path within the reduced-dimensional space.

A candidate mechanistic account follows from the general principle that coordination (synergy) is more fragile than overlap (redundancy) under system perturbation. As a system approaches a critical transition, the effective dimensionality of information transfer decreases. This compression preferentially destroys synergy (the S component of $I = R \times S$), such that integration collapses while redundancy (and thus differentiation) may remain elevated. Under this account, the system moves from the balance zone into the cascade quadrant not by a symmetric process but by the asymmetric fragility of synergy—a pattern consistent with both information-theoretic expectations and the directional dynamics data presented here.

The strong I–D anticorrelation ($\rho = -0.985$ in EEG; Section 3.6.4) means that the proxy metrics operate on a near-one-dimensional manifold parameterized by multi-scale entropy. The four-state model (Table 1) is a theoretical taxonomy of the full I–D plane, not a description of the region accessible with these proxies. CASCADE and RECOVERY dominate empirically because they correspond to movement along the entropy axis; COINCIDENCE and COLLAPSE are rare because they require orthogonal movement that entropy-coupled metrics cannot produce. Future work with entropy-decoupled metrics—partial information decomposition [31], network mutual information, or graph-theoretic topology measures—could test whether the full two-dimensional I–D plane is accessible when integration and differentiation are measured independently. For now, the framework is most accurately described as a multi-scale entropy coordinate system in which critical transitions manifest as rapid, high-variance departures from baseline entropy.

4.2 Critical Slowing Down in I–D Coordinates

The early warning signal (Test 3)—rising variance of $|I - D|$ before events—connects the DET framework directly to the well-established critical transitions literature [1, 2]. Critical slowing down predicts that as a system approaches a bifurcation point, its recovery rate from perturbations decreases, leading to increasing variance and autocorrelation in state variables. The I–D balance metric provides a natural coordinate in which to measure this phenomenon: the system's ability to maintain balance degrades as it approaches the transition, producing larger and larger fluctuations in $|I - D|$.

The quantitative consistency across domains—a 2.0× variance elevation at 5-day lead in both financial and space weather data—is noteworthy. While the temporal scales of the underlying processes differ (financial cascades build over weeks, geomagnetic storms over days), the early warning signal has similar magnitude at similar lead times. This suggests that the I–D coordinate system may capture a domain-general feature of approaching transitions, though the author cautions against over-interpreting numerical coincidences from two domains.

The decay profile of the early warning signal differs between domains in a physically sensible way. Financial markets show persistent elevation out to 30 days, consistent with the slow build-up of systemic stress through leverage accumulation and correlation tightening. Space weather shows rapid decay by 15 days, consistent with the faster dynamics of solar wind–magnetosphere coupling. An I–D framework that showed identical decay profiles across domains would be suspicious; the observed domain-appropriate decay profiles increase confidence that the framework is capturing genuine physics rather than statistical artifacts.

4.3 Why Sum Fails: Structure Versus Magnitude

The EEG results provide the strongest evidence for the DET framework's construct validity. The Sum anti-prediction (AUC 0.288–0.416) is not merely a failure—it is an *informative* failure that illuminates what the framework measures.

Consider two hypothetical explanations for why I–D balance discriminates system states:

1. **Magnitude hypothesis:** Events and state changes are simply associated with higher (or lower) total activity, and the I–D framework captures this through its integration metric.

2. **Structure hypothesis:** Events and state changes are associated with changes in the *organizational structure* of multi-component coupling, which is partially but imperfectly correlated with total activity.

In financial markets and space weather, these hypotheses make similar predictions: cascades involve elevated activity across most layers, so both Sum and I should discriminate. The data are consistent with both hypotheses in these domains (Sum AUC 0.868–0.932, I-based AUC 0.692–0.843).

The EEG domain dissociates the two hypotheses. N3 deep sleep has *higher* total spectral power than wakefulness (massive delta oscillations), so the magnitude hypothesis predicts that Sum should classify N3 as the "event" state—which is indeed what happens, producing the anti-prediction. The structure hypothesis predicts that I $(R \times S)$ should correctly identify wakefulness as the state with higher integration, because wake involves coordinated multi-band activity while N3 involves single-band dominance—which is also what happens (AUC 0.869–0.904).

The dissociation is clean, the effect is large, and it scales with sample size. This constitutes strong evidence that the I–D framework captures organizational structure rather than mere magnitude, validating the construct at the heart of the Dynamic Existence Threshold.

The Φ conservation finding (Test 6) deepens the Sum vs. $R \times S$ distinction. Sum cannot detect component redistribution because it conflates I and D into a single magnitude. The I–D decomposition is necessary precisely because Φ is conserved during transitions—you need to see the components moving independently to detect the transition. A metric that captures only total organizational information (whether as Sum or as Φ directly) would be blind to the very redistribution that defines the critical transition.

A natural question is whether domain-specific spectral ratios—particularly the delta/beta power ratio, a standard sleep staging feature—outperform $R \times S$ on the EEG task. Head-to-head comparison on the same 50-subject cohort yields delta/beta AUC $= 0.912$ vs. $R \times S$ AUC $= 0.901$ (bootstrap 95% CI for the difference: $[-0.013, -0.008]$, $p < 0.05$). The two metrics are highly correlated (Spearman $\rho = 0.923$), confirming that $R \times S$ partially recapitulates spectral power distribution information. However, on harder discrimination tasks—N1 vs. N2 ($R \times S$ AUC 0.740 vs. delta/beta 0.685) and Wake vs. N2 (0.532 vs. 0.502)—$R \times S$ outperforms the spectral ratio, suggesting that the multi-band integration measure captures organizational structure that a two-band ratio misses. Crucially, the delta/beta ratio is undefined outside neuroscience: financial markets and space weather have no frequency bands. The DET framework's contribution is not that $R \times S$ is the optimal classifier for any single domain, but that it provides a substrate-independent coordinate system in which state transitions are interpretable across all three domains tested—including the one where a domain-specific shortcut exists.

4.4 Cross-Domain Consistency and the Limits of Broad Generality

The results demonstrate broad generality of the DET framework across three substrates (socioeconomic, geophysical, biological) without parameter tuning. However, the author is careful to distinguish between two claims:

1. **Narrow claim (supported):** The I–D balance framework provides a useful coordinate system for characterizing critical transitions across multiple domains, and its predictions are confirmed in all three domains tested.

2. **Strong claim (not yet supported):** The I–D balance framework captures a universal principle that applies to all complex systems undergoing critical transitions.

The narrow claim is well-supported by the data. The strong claim would require testing across a much wider range of systems—ecological tipping points, cellular state transitions, social network cascades, engineered system failures—and would need to account for systems where the five-layer decomposition may not have a natural interpretation.

The cross-domain transfer results (Test 4) illustrate both the promise and the limitations of broad generality. Sum-based thresholds transfer reasonably between financial and space weather domains ($1.47\times$ ratio, 0.78 TPR), but $R{\times}S$ thresholds diverge more ($3.77\times$ ratio). This suggests that while the *qualitative* structure of I–D space is domain-general, the *quantitative* calibration of integration metrics is domain-specific. The z-score normalization of the balance metric partially addresses this by converting domain-specific scales to comparable units, but it does not fully eliminate the issue.

Two untested domains merit particular attention. First, artificial neural networks represent a natural candidate for extension: the five-layer architecture maps onto network depth or attention-head groupings, and the Sum anti-prediction result raises the empirical question of whether large-scale AI systems exhibit structured I–D coupling or merely high-magnitude activation—the same question the EEG analysis resolved for biological neural systems. Second, engineered distributed systems (microservice architectures, monitoring infrastructures, sensor networks) already decompose into multi-layer telemetry that maps directly onto the five-layer framework. Service degradation cascades—in which correlated failure propagates across layers while the system's total activity remains high—are structurally analogous to the financial cascades tested here. Both domains are architecturally compatible with the framework but remain empirically untested; the claims of this paper are limited to the three domains for which data are presented.

4.5   Relation to Existing Frameworks

The DET framework intersects with several established theoretical traditions.

Integrated Information Theory [3]. IIT's Φ quantifies integrated information as the information generated by a system above and beyond its parts. The DET framework draws inspiration from IIT's emphasis on integration but differs in two ways: (a) it adds an explicit differentiation coordinate, creating a two-dimensional state space rather than a single scalar; and (b) it uses empirically tractable proxy metrics ($R \times S$ for integration, JSD for differentiation) rather than computing Φ directly, which remains intractable for systems of even moderate size. The four-state model can be seen as a coarse-graining of the I–D plane that may correspond heuristically to regimes of high and low Φ, though $R \times S$ is a computationally tractable proxy rather than a formal information-theoretic measure of integrated information.

Critical transitions theory [1, 27]. The DET framework provides a specific coordinate system (I–D space) and a specific indicator (variance of $B = |z(I) - z(D)|$) for the general phenomenon of critical slowing down. This is complementary to existing approaches that use variance and autocorrelation of raw state variables [2]. The I–D coordinate system has the advantage of being explicitly multi-scale (through the five-layer architecture) and of separating structural organization from magnitude.

Perturbational complexity [24, 28]. The perturbational complexity index (PCI) measures the information content of the brain's response to transcranial magnetic stimulation, and has been shown to discriminate conscious from unconscious states. PCI is high when the response is both integrated (spatially widespread) and differentiated (temporally complex)—precisely the conditions that define the balance zone in the DET framework. The EEG results reported here are consistent with PCI findings, achieved through passive observation rather than perturbation.

Self-organized criticality (SOC) [29, 30, 27]. The balance zone may correspond to the critical state in SOC models—the state to which the system self-organizes and from which avalanches (events) depart. The residence time results (Test 2) are consistent with this interpretation: the system spends more time in the balance zone during normal periods, as expected if the balance zone is an attractor, and exits rapidly during events, as expected if events are excursions from the critical state.

Partial information decomposition (PID) [31, 32, 33]. The R and S metrics are inspired by but not identical to the redundancy and synergy of PID. PID provides exact decompositions of mutual information into redundant, synergistic, and unique components; R and S are empirical proxies designed for tractability across domains. A formal connection between the DET framework's $R \times S$ and PID's synergy would strengthen the theoretical foundations of this work.

4.6 Limitations

Several limitations warrant explicit acknowledgment.

Five-layer architecture. The choice of exactly five layers is pragmatic rather than principled. However, the layer sensitivity analysis (Section 3.7, Table 7) demonstrates that the framework is robust to this choice: adjacent layer counts ($N = 4$ and $N = 6$) produce AUC within 10% of the $N = 5$ baseline across all three domains, with $N = 6$ actually improving performance in EEG (0.924 vs. 0.869) and space weather (0.858 vs. 0.835). The five-layer count is not a local optimum but a pragmatic choice within a broad robustness plateau spanning $N = 4$ to $N = 8$. Performance degrades only at $N = 3$, where insufficient hierarchical depth limits integration measurement.

Proxy metrics and entropy coupling. $R \times S$ and JSD are proxies for integration and differentiation, not exact information-theoretic quantities. Critically, both are functions of the Shannon entropy H of the layer distribution: R is derived from $N_{\text{eff}} = \exp(H)$ and increases with H, while $D = \mathrm{JSD}^{1/2}(\mathbf{p}\|\mathbf{u})$ decreases with H as the distribution approaches uniformity. This entropy coupling produces a Spearman anticorrelation of $\rho = -0.985$ in EEG (Section 3.6.4), constraining the nominally two-dimensional I–D space to a near-one-dimensional manifold. Two of the four theoretical states (Organized complexity: high I, high D; Death/off: low I, low D) are rendered empirically inaccessible because they require I and D to covary rather than anti-covary. Under near-perfect anticorrelation, the balance metric $B = |z(I) - z(D)|$ reduces to approximately $|2\,z(I)|$—the absolute deviation of entropy from its historical mean. Entropy-decoupled measures—such as network mutual information, PID synergy [31], or graph-theoretic connectivity—might yield a genuinely two-dimensional I–D space in which all four states are accessible. The empirical findings reported here (cross-domain transfer, early warning signals, structure-versus-magnitude dissociation) do not depend on the dimensionality claim and hold on the entropy axis alone.

Temporal resolution. Financial and space weather data are analyzed at daily resolution, while EEG is analyzed at 30-second epoch resolution. The framework has not been tested at intermediate timescales or at the sub-second resolution relevant to neural dynamics. The balance zone dynamics might differ qualitatively at different temporal scales.

Event definition. The definition of "events" in each domain relies on established criteria (VIX thresholds, Dst thresholds, sleep stage scoring), but these criteria are themselves imperfect. The framework's performance could be sensitive to event definition, particularly for boundary cases.

Sample size in EEG. The current results are based on 50 subjects (57,212 epochs for Tests 1–5; 136,394 epochs across all stages for Test 6). While the signal strengthens monotonically from 5 to 50 subjects (Table 4), the sample remains modest by neuroimaging standards. Replication in larger and more diverse cohorts is needed.

Causal ambiguity. The framework demonstrates that I–D balance *discriminates* system states and that its variance provides *early warning* of transitions, but it does not establish that I–D imbalance *causes* transitions. The relationship could be correlational, with both I–D dynamics and transitions driven by a common underlying process.

Domain selection. Three domains, while spanning physical, biological, and socioeconomic systems, constitute a limited sample. Domains were selected in part because they have well-defined "event" periods and accessible multi-layer data—a convenience sample that may not be representative of complex systems in general.

Approximate conservation. The Φ conservation (Test 6) is approximate, not exact. The 14% deviation at the N2→N3 transition and the statistically significant (though small-magnitude) deviation in financial markets indicate that Φ is not a strict conserved quantity. The consistency test distinguishes conservation from destruction (critical transitions vs. dimensional embedding), but finer questions—whether the residual deviations reflect genuine non-conservation, measurement noise, or proxy metric limitations—remain open and would benefit from entropy-decoupled metrics in future work.

5 Conclusion

This paper introduced the Dynamic Existence Threshold—the hypothesis that systems persist in the zone where integration and differentiation are jointly sustained, and that critical events correspond to departures from this zone. Six quantitative tests, validated by permutation-based negative controls and comparison with standard early warning indicators, across three domains (financial markets, space weather, and human EEG) support this hypothesis:

1. Events cluster at I–D extremes while normal states occupy the balance center, with the CASCADE pattern (rising I, falling D) accounting for 45–51% of event dynamics.

2. Events exit the balance zone faster than normal state changes, with residence times halved during event periods ($p = 0.0001$ in financial data).

3. Rising variance of the balance metric $B = |z(I) - z(D)|$ provides early warning 5–30 days before events, with a 2.0× elevation ratio observed independently in both financial and space weather domains.

4. Exit velocity discriminates events from normal states with AUC 0.692–0.843, and cross-domain threshold transfer achieves 0.78 TPR without recalibration.

5. In EEG, the coupling structure metric $R \times S$ achieves AUC 0.909 (95% CI [0.904, 0.913]) for discriminating wakefulness from deep sleep, while the simple

sum anti-predicts (AUC 0.288–0.416), confirming that the framework captures organizational structure rather than magnitude.

6. Total organizational information $\Phi = I + D$ is approximately conserved during critical transitions across all three domains (within 1–14%), with integration and differentiation undergoing large inverse changes. Financial cascades and space weather storms show the CASCADE pattern (I surges, D collapses), while EEG sleep transitions show the DISSOLVE pattern (I collapses, D surges). Dimensional embedding, by contrast, destroys Φ irreversibly ($\sim$86% loss), identifying two distinct failure modes for organized systems: reversible information redistribution versus irreversible information destruction.

These results extend integrated information theory and critical transitions research into a dynamic framework that characterizes not just whether a system can persist as an organized entity, but the conditions under which it transitions between qualitatively different states. The I–D balance zone emerges as a broadly general feature of complex systems, providing a common coordinate system for understanding critical transitions across substrates. The Φ conservation property connects this dynamic framework to the static existence threshold [34]: systems do not lose organizational information during critical transitions—they redistribute it. The DET framework detects this redistribution by decomposing Φ into its integration and differentiation components, revealing the structural shift that total Φ alone would hide.

The practical implication is a domain-general early warning indicator: rising variance of the balance metric B, computed from any system that can be decomposed into multiple measurement layers. The theoretical implication is that the multi-scale entropy captured by the integration–differentiation decomposition provides a coordinate system in which the approach to critical transitions is measurable, with the balance zone corresponding to entropy near its historical baseline and departures signaling organizational state change.

Supplementary Material

See supplementary material for complete analysis code, pre-computed results, and publication figures at https://github.com/existencethreshold/dynamic-existence-threshold.

Acknowledgments

The framework emerged from practical work on distributed system monitoring, which motivated the integration-differentiation decomposition.

AI tools acknowledgment: AI tools (Claude, Gemini) were used to assist with code generation, statistical analysis, and manuscript revision. All scientific claims, interpretations, and editorial decisions are the sole responsibility of the author.

Author Declarations

Conflict of Interest: The author has no conflicts to disclose.

Ethics Approval: This study analyzed publicly available, de-identified datasets. The EEG data arc from the Sleep-EDF dataset hosted on PhysioNet; ethical approval for the original data collection was obtained from the dataset creators (Kemp et al., 2000).

Author Contributions (CRediT): Nathan M. Thornhill: Conceptualization (lead); Data curation (lead); Formal analysis (lead); Investigation (lead); Methodology (lead); Software (lead); Validation (lead); Visualization (lead); Writing – original draft (lead); Writing – review & editing (lead).

Data Availability Statement

The data that support the findings of this study are openly available. Financial data are derived from publicly available market indices via the yfinance Python library. Space weather data are from NASA OMNI2 (https://omniweb.gsfc.nasa.gov/). EEG data are from the Sleep-EDF dataset on PhysioNet (https://physionet.org/content/sleep-edfx/). Complete analysis code and pre-computed results are available at https://github.com/existencethreshold/dynamic-existence-threshold.

References

[1] M. Scheffer, J. Bascompte, W. A. Brock, V. Brovkin, S. R. Carpenter, V. Dakos, H. Held, E. H. van Nes, M. Rietkerk, and G. Sugihara, "Early-warning signals for critical transitions," Nature **461**(7260), 53–59 (2009).

[2] V. Dakos, S. R. Carpenter, W. A. Brock, A. M. Ellison, V. Guttal, A. R. Ives, S. Kéfi, V. Livina, D. A. Seekell, E. H. van Nes, and M. Scheffer, "Methods for detecting early warnings of critical transitions in time series illustrated using simulated ecological data," PLoS ONE **7**(7), e41010 (2012).

[3] G. Tononi, M. Boly, M. Massimini, and C. Koch, "Integrated information theory: From consciousness to its physical substrate," Nat. Rev. Neurosci. **17**(7), 450–461 (2016).

[4] G. Tononi, "An information integration theory of consciousness," BMC Neurosci. **5**, 42 (2004).

[5] B. J. Baars, *A Cognitive Theory of Consciousness* (Cambridge University Press, 1988).

[6] A. K. Seth, A. B. Barrett, and L. Barnett, "Causal density and integrated information as measures of conscious level," Philos. Trans. R. Soc. A **369**(1952), 3748–3767 (2011).

[7] R. B. Berry, R. Brooks, C. E. Gamaldo, S. M. Harding, R. M. Lloyd, C. L. Marcus, and B. V. Vaughn, "The AASM manual for the scoring of sleep and associated events: Rules, terminology and technical specifications, version 2.4," American Academy of Sleep Medicine (2017).

[8] A. V. Oppenheim and A. S. Willsky, *Signals and Systems*, 2nd ed. (Prentice Hall, 1997).

[9] S. D. Muthukumaraswamy, "High-frequency brain activity and muscle artifacts in MEG/EEG: A review and recommendations," Front. Hum. Neurosci. **7**, 138 (2013).

[10] N. E. Papitashvili and J. H. King, "OMNI hourly data set," NASA Space Physics Data Facility (2020). https://omniweb.gsfc.nasa.gov/

[11] K. J. Forbes and R. Rigobon, "No contagion, only interdependence: Measuring stock market comovements," J. Finance **57**(5), 2223–2261 (2002).

[12] M. O. Hill, "Diversity and evenness: A unifying notation and its consequences," Ecology **54**(2), 427–432 (1973).

[13] L. Jost, "Entropy and diversity," Oikos **113**(2), 363–375 (2006).

[14] E. C. Pielou, "The measurement of diversity in different types of biological collections," J. Theor. Biol. **13**, 131–144 (1966).

[15] J. Lin, "Divergence measures based on the Shannon entropy," IEEE Trans. Inf. Theory **37**(1), 145–151 (1991).

[16] D. M. Endres and J. E. Schindelin, "A new metric for probability distributions," IEEE Trans. Inf. Theory **49**(7), 1858–1860 (2003).

[17] W. D. Gonzalez, J. A. Joselyn, Y. Kamide, H. W. Kroehl, G. Rostoker, B. T. Tsurutani, and V. M. Vasyliunas, "What is a geomagnetic storm?" J. Geophys. Res. **99**(A4), 5771–5792 (1994).

[18] B. Kemp, A. H. Zwinderman, B. Tuk, H. A. C. Kamphuisen, and J. J. L. Oberyé, "Analysis of a sleep-dependent neuronal feedback loop: The sleep-EDF database," IEEE Trans. Biomed. Eng. **47**(9), 1185–1194 (2000).

[19] A. L. Goldberger, L. A. N. Amaral, L. Glass, J. M. Hausdorff, P. Ch. Ivanov, R. G. Mark, J. E. Mietus, G. B. Moody, C.-K. Peng, and H. E. Stanley, "PhysioBank, PhysioToolkit, and PhysioNet: Components of a new research resource for complex physiologic signals," Circulation **101**(23), e215–e220 (2000).

[20] Y. Benjamini and Y. Hochberg, "Controlling the false discovery rate: A practical and powerful approach to multiple testing," J. R. Stat. Soc. Ser. B **57**(1), 289–300 (1995).

[21] M. Massimini, F. Ferrarelli, R. Huber, S. K. Esser, H. Singh, and G. Tononi, "Breakdown of cortical effective connectivity during sleep," Science **309**(5744), 2228–2232 (2005).

[22] E. Niedermeyer and F. L. da Silva, *Electroencephalography: Basic Principles, Clinical Applications, and Related Fields*, 5th ed. (Lippincott Williams & Wilkins, 2005).

[23] G. Buzsáki, *Rhythms of the Brain* (Oxford University Press, 2006).

[24] A. G. Casali, O. Gosseries, M. Rosanova, M. Boly, S. Sarasso, K. R. Casali, S. Casarotto, M.-A. Bruno, S. Laureys, G. Tononi, and M. Massimini, "A theoretically based index of consciousness independent of sensory processing and behavior," Sci. Transl. Med. **5**(198), 198ra105 (2013).

[25] F. Amzica and M. Steriade, "Electrophysiological correlates of sleep delta waves," Electroencephalogr. Clin. Neurophysiol. **107**(2), 69–83 (1998).

[26] K. Lehnertz, S. Bialonski, M.-T. Horstmann, D. Krug, A. Rothkegel, M. Staniek, and T. Wagner, "Synchronization phenomena in human epileptic brain networks," J. Neurosci. Methods **183**(1), 42–48 (2009).

[27] S. H. Strogatz, *Nonlinear Dynamics and Chaos*, 2nd ed. (Westview Press, 2015).

[28] S. Sarasso, M. Boly, M. Napolitani, O. Gosseries, V. Charland-Verville, S. Casarotto, M. Rosanova, A. G. Casali, J.-F. Brichant, P. Boveroux, S. Rex, G. Tononi, S. Laureys, and M. Massimini, "Consciousness and complexity during unresponsiveness induced by propofol, xenon, and ketamine," Curr. Biol. **25**(23), 3099–3105 (2015).

[29] P. Bak, C. Tang, and K. Wiesenfeld, "Self-organized criticality: An explanation of $1/f$ noise," Phys. Rev. Lett. **59**(4), 381–384 (1987).

[30] Z. Olami, H. J. S. Feder, and K. Christensen, "Self-organized criticality in a continuous, nonconservative cellular automaton modeling earthquakes," Phys. Rev. Lett. **68**(8), 1244–1247 (1992).

[31] P. L. Williams and R. D. Beer, "Nonnegative decomposition of multivariate information," arXiv preprint arXiv:1004.2515 (2010).

[32] M. Aguilera, S. A. Moosavi, and H. Shimazaki, "Information-theoretic complexity and the structure of large-scale neural systems," arXiv preprint arXiv:1906.04035 (2019).

[33] N. J. Popiel, S. Bhatt, M. Engström, K. Lindgren, and J. T. Lizier, "Partial information decomposition and the emergence of redundancy in complex systems," Phys. Rev. E **102**(3), 032110 (2020).

[34] N. M. Thornhill, "The existence threshold: A framework for pattern persistence in binary discrete systems," Zenodo (2026). https://doi.org/10.5281/zenodo.18124074

[35] N. M. Thornhill, "Pattern loss at dimensional boundaries: The 86% scaling law," Zenodo (2026). https://doi.org/10.5281/zenodo.18262424

[36] N. M. Thornhill, "The dimensional loss theorem: Proof and neural network validation," Zenodo (2026). https://doi.org/10.5281/zenodo.18319430

INSTITUTE FOR COMPLEXITY SCIENCE AND ADVANCED COMPUTING

ABOUT THE AUTHOR

NATHAN M. THORNHILL

Nathan M. Thornhill is a complexity science researcher working in information theory and pattern persistence. The four papers collected here develop a single instrument — a scalar metric for pattern persistence and the cost of dimensional change — validated across cellular automata, transformer attention, financial markets, space weather, and the human brain in sleep and wake.

He came to this field from outside the traditional academy. Years in healthcare administration — nursing homes first, then ICU admissions — put him across from coordinated systems succeeding, failing, and rearranging themselves on a nightly basis. Computers and technology have been a lifelong passion; AI, when it arrived, read not as the latest instrument but as the most natural subject the research could be pointed at. The four papers come out of the three threads running together.

He holds provisional patents on measurement systems derived from this work, with applications in artificial intelligence, artificial general intelligence, and AI/AGI safety. His research is archived through the Institute for Complexity Science and Advanced Computing.

He lives in Fort Wayne, Indiana, where he also runs a technology consultancy. He enjoys aviation and flight, camping, plays guitar, and spends too much time trying to maintain a fairly acceptable lawn and landscaping, but family always comes first. The lab may just be a spare room, but the hours — they are late, and the work must continue.

nathanthornhill.com · icsacinstitute.org · info@icsacinstitute.org
ORCID 0009-0009-3161-528X

Institute for Complexity Science and Advanced Computing

Colophon

Title

Foundations of the Existence Threshold: The Scholarly Collection

Author

Nathan M. Thornhill
Fort Wayne, Indiana, USA

Publisher

Institute for Complexity Science and Advanced Computing
Fort Wayne, Indiana, USA

Editorial Status. These four papers are preprints that have been read and scored by the Institute's open-review pipeline — advanced artificial intelligence working in collaboration with human reviewers. The panel review, the artificial intelligence accuracy and quality audit, and current editorial status are published per paper at icsacinstitute.org/publications.

Permissions & Contact. Direct editorial correspondence and permission requests to info@icsacinstitute.org. Verify Institute affiliation at icsacinstitute.org/verify.

INSTITUTE FOR COMPLEXITY SCIENCE AND ADVANCED COMPUTING

An independent research organization advancing the formal study of complexity science: pattern persistence, dimensional dynamics, and the computational systems in which they apply. For generations, the scholarly record has sat behind paywalls, closed reviews, and article-processing charges that bill authors thousands for the privilege of publishing. ICSAC was built to end that. Every paper arrives with the full editorial trail attached — preprint, panel reviews, and quality-control records, in their entirety. Submissions are judged on the science, not the author's affiliation or lack of one. Peer review is performed by a collaboration of artificial intelligence and human oversight, on infrastructure the Institute builds and runs. What legacy publishing takes a year to deliver, ICSAC delivers in weeks, and the record stays permanently open and unbiased.